THE BHAKTIVEDANTA INSTITUTE MONOGRAPH SERIES

THE BHAKTIVEDANTA INSTITUTE MONOGRAPH SERIES

Produced for the
1977 "Life Comes from Life" Conference
– Vrindavan, India –

Thoudam D. Singh, Ph.D.
and Richard L. Thompson, Ph.D.

Edited by S. E. Kreitzer, Ph.D.

IVS

INSTITUTE FOR VAISHNAVA STUDIES
Gainesville, Florida

Contact information:
www.richardlthompson.com
rlthompsonarchives@ivs.edu

Write to:
Richard L. Thompson Archives
12834 NW 151 Road
Alachua, Florida 32615

ISBN: 978-0-9981871-7-4

Cover art: ©iStock.com | ClaudioVentrella

Cover description: *Escherichia coli*, a unicellular organism about 500 times smaller than the average cells of higher plants and animals, is one of the smallest and simplest of all living organisms. Yet it is estimated that a single *E. coli* cell contains between 3,000 and 6,000 different types of molecules. Among these is a DNA molecule containing coded instructions for the construction of all the other molecules.

According to James Watson, one of the foremost authorities on molecular biology, "We must immediately admit that the structure of a cell will never be understood in the same way as that of water or glucose molecules. Not only will the exact structures of most macromolecules remain unsolved, but their relative location within cells can be only vaguely know." (p.117)

To Bhaktisvarūpa Dāmodara Mahārāja
and Sadāpūta Dāsa, who devoted much of their lives
to pioneering the Bhaktivedanta Institute (BI).

In addition, a special appreciation to Yamarāja Dāsa,
who has served for more than four decades as graphic designer for
Back to Godhead magazine, the publisher of numerous BI essays
intended for a wider readership. Yamarāja also designed
Richard L. Thompson's books, as well as the works for many other
authors too numerous to count, all exploring the mysteries of
the Gaudiya Vaishnava tradition within a contemporary context.

*From left: Svarūpa Dāmodara Dāsa (Thoudam D. Singh), Sadāpūta Dāsa
(Richard L. Thompson), Aniruddha Dāsa, His Divine Grace A. C. Bhaktivedanta Swami
Prabhupāda, Rupānuga Dāsa (Robert Corens), Mādhava Dāsa (Michael Marchetti).
Photo taken in Washington, D. C., July 1976, at the ISKCON center.*

CONTENTS

PREFACE

In October 1977, scientist disciples of A. C. Bhaktivedanta Swami Prabhupāda gathered at the new Krishna Balaram Temple in Vrindavan, India, for a conference. Vrindavan is one of the holiest places in India with possibly the highest density of temples of any town in the world. In this exotic setting, Prabhupāda proposed a conference entitled "Life Comes from Life" to argue the case that life does not come from matter as proposed by popular theories. The attendees presenting at the conference included disciples of Prabhupāda with a scientific background, as well as distinguished scientists and philosophers expressing various points of view. Prabhupāda's intention was for there to be both strong argument and dialogue. Toward that end, the monographs reproduced in this volume played an important role.

My story for this conference also began in India. I was a Peace Corps Volunteer from 1971–72, assigned to teach science at a rural high school in Bihar, India. While visiting Kolkata, I met disciples of Prabhupāda who invited me to assist in food relief for Bangladesh refugees in rural West Bengal. While there, I had the good fortune to meet Śrīla Prabhupāda as his only outside guest at a small festival commissioned on his society's recently purchased land in Mayapur. I also enjoyed a personal three-day conversation with Prabhupāda in a hut by the Ganges River, where we discussed science, philosophy and other topics. Later, Prabhupāda requested that the recordings of our conversation be published into a book that he titled *Perfect Questions, Perfect Answers* (now published in 40 languages).

Fast forward to 1977. The authors of the monographs featured in this volume, Dr. Singh and Dr. Thompson, approached me about participating at the "Life Comes from Life" conference being organized in Vrindavan, using my science background in geology to present a paper entitled "Darwin's Theory and the Past History of Life." This was also an opportunity for me to see Prabhupāda one last time, before he passed away on November 14, 1977.

The purpose of the Bhaktivedanta Institute (BI), which organized the conference, was featured on the inside cover of their journal, *Perspectives on Bhaktivedanta Institute* (1979), in a description of Prabhupāda's intentions:

> Prabhupāda was never one to criticize modern, materialistic science per se, but only insofar as it ignores and rejects the fully verifiable findings of the higher science of transcendence – of which he was an unquestioned master . . . [He] was especially critical of the "life-comes-from-matter" theory . . . [and thus] charged the Bhaktivedanta Institute with the primary task of disproving this through the use of the most sophisticated techniques of modern mathematics, chemistry, and physics themselves.

The attendees presented their papers, debated, and developed areas of mutual understanding. Prabhupāda commented to his disciples after the conference, "Bhaktivedanta Institute is doing something scientifically to understand God consciousness." He encouraged his followers to "go on proceeding like [this] more and more. So many scientists, foreign and local, they participated, discussed. It is not ordinary thing. Hmm?" Prabhupāda then offered his conclusion, "the importance of Bhaktivedanta Institute is there . . . 'Come on,' we are challenging. 'Discuss like scientist, not like sentimentalists.'" (Vrindavan, October 27, 1977)

As Prabhupāda was too ill to attend the conference, we scientist disciples gathered in his room each evening to give a report. Prabhupāda wanted to know how each side argued and counter-argued. He was quite pleased with the outcome.

As an aside, I also had a personal connection with the three monographs featured in this edition during their passage to India that proved an adventure all its own. These volumes came off the press in Boston only shortly before the conference began, and I was the last attendee leaving from America, flying out from the Pan Am terminal at New York City's JFK airport. A dear friend in Boston, Kardama Muni Dāsa, received a midnight call from one of the organizers in India with the request, "Get these monographs to us at all costs!" In the rush, he missed the last scheduled flight to JFK. The only available option left was the Eastern Shuttle to LaGuardia Airport followed by a relatively short taxi ride to JFK. But upon landing at LaGuardia, Kardama learned that the airport taxis were blocked – they could not reach JFK due to street protests surrounding the airport challenging Concorde supersonic transport flights and their related noise. At the last minute, Kardama arranged for a helicopter flight from LaGuardia to the Pan Am terminal at JFK, and somehow we connected with the boxes of monographs just as I readied to board my flight for India. It was meant to be!

We provided these monographs produced by the BI to the conference attendees, and have now combined them for contemporary readers in this 40th Anniversary commemorative volume. Although some of the arguments and scientific details have changed over the past several decades, the monographs remain thought-provoking. For me, they retain their relevance by offering many fascinating insights enriching my own journey of understanding, within the noumenal and phenomenal worlds.

Bob Cohen (Brahmatīrtha Dāsa), Director
Bhaktivedanta Institute for Higher Studies
M. S. Geology, B. S. Chemistry
Gainesville, Florida, USA

INTRODUCTION

This edition of facsimile reprints features the three monographs produced for the Bhakti-vedanta Institute's October 14–16, 1977, "Life Comes from Life" conference hosted at ISK-CON's Krishna Balaram Mandir in Vrindavan, India. Included at the end of the volume are related conference documents that, along with the three monographs, provide important examples of the Institute's early attempts at facilitating cross-disciplinary discourse in science, religion, and philosophy. No doubt the monograph's two authors, Thoudam Singh and Richard Thompson, wished to credibly uphold the central tenants of the Gaudiya Vaishnava tradition with which they identified, while also displaying respect for the integrity of the modern disciplines with mature appreciation drawn from experience. Singh earned his Ph.D. in Physical Organic Chemistry at the University of California, Irvine, while subsequently working as a Research Fellow at Emory University. Meanwhile, Thompson earned his Ph.D. in Probability Theory and Statistical Mechanics at Cornell University as a student of Frank Spitzer. Thompson's dissertation, "Equilibrium States on Thin Energy Shells," was chosen for publication in *Memoirs of the American Mathematical Society*, Number 150.

Along with the three monographs presented in Section I, Section II includes the following items related to the conference: (1) the title page from an early cooperative effort that would eventually become the first monograph, (2) an announcement for the "Life Comes from Life" conference offering a brief outline of its objectives, (3) a brochure titled, "Bhaktivedanta Institute Lecture Series," which presents a project mission statement along with an outline of programs, areas of research, and short biographies of the Bhaktivedanta Institute members speaking at the conference, (4) copies of a trifold invitation brochure originally printed on light yellow cardstock offering a description of conference objectives and a schedule of events, (5) a list of speakers with a description of their professional qualifications at the time of the conference, (6) a finalized schedule of events with the name and topic of each presenter and, (7) text from an article reporting on the conference published in *The Statesman*, a widely distributed national newspaper headquartered in Kolkata. This account is taken from audio recordings of the article being read during a follow-up meeting to the conference. The final document (8) was published in the December 1977 edition of *Back to Godhead*, "the magazine of the Hare Krishna Movement," reporting the events to ISKCON's congregation.

The publishers of the original Monograph Series received a short review written by Professor of Physics at Washington State University, James L. Park, for the second monograph of the series. Dr. Park's comments perhaps offer a fair assessment of the academic relevancy of these publications:

> This refreshing and provocative monograph features a very competent theoretical assault upon the pervasive reductionism so characteristic of present-day thought in biochemical orthodoxy.

The thesis is therefore controversial; but the presentation is, from a standpoint of theoretical physics, both rational and substantive. The work does to be sure defy "normal" scientific attitudes, and it may of course be wrong. However, the cogency of the arguments will surely prevent any reasonable scholar from regarding the work as crank literature merely because of its unconventionality.

On the fortieth anniversary of the conference, now these works are available for consideration by a fresh generation of scholars.

S. E. Kreitzer, Editor
Ph.D. History of Science
October 14, 2017

SECTION I

THE BHAKTIVEDANTA INSTITUTE MONOGRAPH SERIES

Monograph 1

WHAT IS MATTER AND WHAT IS LIFE?

The Bhaktivedanta Institute
Monograph Series Number 1

WHAT IS MATTER

AND

WHAT IS LIFE?

by
Thoudam D. Singh, Ph.D.
(Svarūpa Dāmodara Dāsa Brahmacārī)
Richard L. Thompson, Ph.D.
(Sadāputa Dāsa Adhikārī)

WHAT IS MATTER

AND

WHAT IS LIFE?

The Bhaktivedanta Institute Monograph Series:

Number 1.   What is Matter and What is Life?

Number 2.   Demonstration by Information Theory that Life Cannot Arise from Matter

Number 3.   Consciousness and the Laws of Nature

Information regarding these monographs is available upon request from Bhaktivedanta Institute at:

70 Commonwealth Avenue	Hare Krishna Land
Boston, Massachusetts 02116	Juhu, Bombay 400 054
U.S.A.	India

The Bhaktivedanta Institute
Monograph Series Number 1

What is Matter

and

What is Life?*

by
Thoudam D. Singh, Ph.D.
(Svarūpa Dāmodara Dāsa Brahmacārī)
Richard L. Thompson, Ph.D.
(Sadāputa Dāsa Adhikārī)

*This monograph forms part of a forthcoming book, *The Origin of Life and Matter*, by Thoudam D. Singh, Michael Marchetti, and Richard L. Thompson.

Published by: Bhaktivedanta Institute
                  Boston • Bombay

iii

Readers interested in the subject matter of this monograph are invited to send correspondence to the Bhaktivedanta Institute at the following addresses.

70 Commonwealth Avenue
Boston, Massachusetts 02116
U.S.A. (617) 266-8369

Hare Krishna Land
Juhu, Bombay 400 054
India (Phone: 57-9373)

Bhaktivedanta Gurukula and
Institute for Higher Studies
Bhaktivedanta Swami Marg
Vrindavana, Mathura
India

© 1977 Bhaktivedanta Book Trust

Printed in the United States of America

Library of Congress Catalogue Card Number: 77-89121

Dedicated to His Divine Grace
A. C. Bhaktivedanta Swami Prabhupāda
Who so kindly taught us that the goal of all
scientific knowledge is to understand:
kṛṣṇas tu bhagavān svayam
(*Śrīmad Bhāgavatam* 1.3.28)

About Bhaktivedanta Institute

Bhaktivedanta Institute is a center for advanced study and research into the Vedic scientific knowledge concerning the nature of consciousness and the self. The Institute is the academic division of the International Society for Krishna Consciousness. It consists of a body of scientists and scholars who have recognized the unique value of the teachings of Krishna Consciousness brought to the West by His Divine Grace A. C. Bhaktivedanta Swami Prabhupāda. The main purpose of the Institute is to explore the implications of the Vedic knowledge as it bears on all features of human culture, and to present its findings in courses, lectures, monographs, books, and a quarterly journal, *Sa-vijñānam*.

The Institute presents modern science and other fields of knowledge in the light of Vaiṣṇava philosophy and tradition, providing a new perspective on reality quite different from that of our modern educational systems. One reason for the increasing interest of modern intellectuals in Śrīla Prabhupāda's teachings is doubtlessly the growing awareness that in spite of great scientific and technological advancements, the real goal of human life has somehow been missed. The philosophy of Bhaktivedanta Institute provides a meaningful answer to this concern by proposing that life—not matter—is the basis of the world we perceive.

The central doctrine of modern science is that all phenomena, including those of life and consciousness, can be fully explained and understood by recourse to matter alone. The dictum that "life is a manifestation of matter" is, indeed, the ultimate rationale for the entire civilization of material aggrandizement. The Vedas, on the other hand, teach that conscious life is original, fundamental, and eternal. This is the essence of *Bhagavad-gītā*—"*ahaṁ sarvasya prabhavo mattaḥ sarvaṁ pravartate.*" (10.8) On this fundamental and critical point, modern science and Vedic knowledge find themselves opposed.

Bhaktivedanta Institute is dedicated to disseminating this most fundamental knowledge throughout the world. The Institute is clearly demonstrating that the Vedic version is not a matter simply of "faith" or "belief", but is scientific in the strict sense of the term. Although many of its features may appear difficult to verify experimentally, others have direct implications concerning what

we may expect to observe. Thus, this view should serve as a stimulating challenge to the truly scientific spirit to go beyond the very restrictive framework imposed on our scientific understanding of nature over the last two hundred years. Modern science began as an experiment to see how far nature could be explained without invoking God. But the purpose of Bhaktivedanta Institute is to introduce Vedic knowledge on a genuinely scientific basis for the first time in the history of this modern scientific age.

CONTENTS

Introduction

kṣetra-kṣetrajñayor jñāñaṁ

To understand the distinction between matter (the field or body)
and life (spirit, or the knower of the field) is called knowledge.
Bhagavad-gītā (13.3)

What is matter and what is life? What are their origins? Although these questions have been pondered time and again by many eminent thinkers in both the scientific and philosophical worlds, they have never been answered to everyone's satisfaction, and in spite of such great intellectual endeavor, they still remain quite controversial. Over the past two centuries, the ever increasing success scientists have experienced in their investigation of gross matter has led many people to expect that life will eventually be explained solely as a phenomenon of matter. At the present time nearly all serious attempts to understand the origin of life have been based on this fundamental presupposition, and this controversy is thus being conducted within very narrow limits.

In recent years, scientists of many disciplines, such as chemistry, biology, biochemistry, biophysics, geology, geochemistry, and space science, have devoted considerable attention to the study of the origin of life.[1-4] Virtually all these studies are based on the assumption that life is a manifestation of matter. Scientists in these areas proclaim that life originated from a random combination of molecules interacting under the influence of blind natural laws over a long span of time. These scientists postulate a primordial chemical soup of small and simple molecules, and they imagine that in the course of time, under the influence of chance and mechanical laws, life generated itself from these molecules.

Such speculations date back at least as far as the time of Darwin, who noted that we could not expect to observe life originating in this way today since already existing living organisms would interfere with the process:

It is often said that all the conditions for the first production of a
living organism are now present which could ever have been present.
But if (and oh what a big if) we could conceive in some warm little
pond with all sorts of ammonia and phosphoric salts, light, heat,

electricity, etc., present, that a protein compound was chemically formed, ready to undergo still more complex changes, such matter would at the present day be instantly devoured, or absorbed, which would not have been the case before living creatures were formed.[5]

In Darwin's brief description we see nearly all the basic features of the "primordial soup" that serves as the starting point of modern theories of life's origin. The basic assumption is that living organisms consist of combinations of a few simple chemical compounds. This leads to the hypothesis that simple natural processes may have brought such compounds together under conditions suitable for their combination into more complex forms.

Once this simple initial condition is assumed, the next step is to introduce "chance." In the words of Jacques Monod:

> Chance alone is at the source of every innovation, of all creation in the biosphere. Pure chance, absolutely free but blind, is at the very root of the stupendous edifice of evolution: this central concept of modern biology is no longer one among other possible or even conceivable hypotheses. It is today the sole conceivable hypothesis, the only one that squares with observed and tested fact. And nothing warrants the supposition—or the hope—that on this score our position is likely ever to be revised.[6]

Chance and the interaction of molecules in accordance with simple physical laws are the only factors admitted as causes of change in modern scientific theories of nature. Although these causes seem inadequate, it is assumed that if a sufficiently long period of time is granted they will be capable of generating life in all its diverse and complex forms. Modern scientific inquiry into the origin of life thus adheres to the basic model illustrated in figure 1.

It is our thesis that this model is based on a fundamental misunderstanding of what life actually is. Before inquiring into the origin of a certain thing, it is essential to understand its fundamental nature. Otherwise, searching for its origin is totally meaningless. In this paper we will therefore attempt to understand the fundamental differences between matter and life before we inquire into the origin of life. We will argue that present scientific knowledge, especially that of physics and chemistry, is unable to fully explain the intricate phenomena associated with life. As the French physicist Louis de Broglie remarked, "It is premature to reduce the vital process to the quite insufficiently developed con-

CHANCE, MOLECULAR FORCES
AND A LONG TIME SPAN.

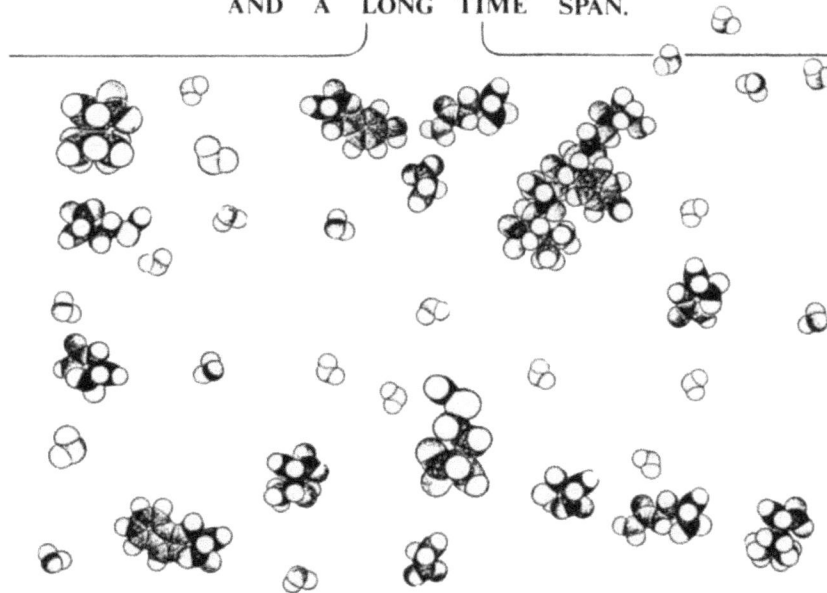

Figure 1. Is this the origin of life?

3

ception of 19th and even 20th century chemistry and physics."[7]

The physical sciences study "gross matter" only, although their results have been extrapolated to explain life. This reductionistic approach, however, has only indicated that the known physical laws of the present day are quite insufficient to account for the features of life. A new paradigm describing life and the laws of nature is needed. We will, therefore, propose an alternative scientific viewpoint (hypothesis). We do not intend to prove anything rigorously in this paper; rather, we will discuss the implications of this alternative viewpoint from a general perspective.

The Basic Features of the Absolute Truth
According to Modern Science

The goal of scientific investigations is to seek the ultimate cause of all phenomena, governing matter, life, and the universe we perceive. In the course of history, many great scientists and philosophers have encountered considerable difficulties in this attempt, and many have acutely felt the limitations of the human intellect. Nonetheless, the search for knowledge is an inherent quality of the inquisitive mind, and it will go on.

In modern science, the concept of the ultimate cause, or the absolute truth, seems to be vaguely incorporated into the physical laws called the laws of nature. According to the theory of evolution, these laws of nature, and nothing else, have the power to select the most suitable forms from among different possibilities. We shall therefore briefly examine what these laws are.

The physical sciences—physics, chemistry, and mathematics—are devoted to the study of matter only. Copernicus, Galileo, Kepler, and Newton in the fifteenth, sixteenth, and seventeenth centuries were pioneers in the study of gross material phenomena, such as the planetary motions. These objects of study were found to obey certain mathematical regularities, which were termed "laws of nature," and the discovery of such laws became the target of scientific investigations. People were greatly impressed by the discovery of Newton's law of gravitation, Kepler's laws of planetary motion, and so on. Thus the more ambitious scientists came to think that all the phenomena underlying nature could be described by simple mathematical equations.

Among the more enthusiastic ones, the French scientist Pierre de Laplace proudly announced in the beginning of the nineteenth century that "all the effects of nature are only the mathematical consequences of a small number of immutable laws."[8] He believed that the universe was made up of atomic particles and that the exact condition of the universe at any one time could be given by specifying the exact positions and velocities of those particles with respect to a system of coordinates. He claimed that given these positions and velocities, he could, at least in principle, calculate the entire past and future of the universe from the laws of motion governing the particles. (Laplace, of course, could not have lived up to his boast, and any honest scientist might have thought that such

5

big statements would not survive the test of time.)

From the art of alchemy in the Middle Ages, chemistry developed, and later it became one of the most significant fields of study in science. Antoine Lavoisier in the eighteenth century and John Dalton in the early part of the nineteenth laid the foundations for modern chemistry. They proposed that all gross substances were made up of elements or atoms, and since then the discovery of atomic elements has played a central role in chemistry. The discovery of the periodic law of chemical elements by the Russian scientist Mendeleev during the nineteenth century greatly impressed chemists and physicists all over the world. With this discovery, as well as with the discovery of many other laws, such as the ideal gas laws, the laws of dilute solutions, and the laws of thermodynamics, scientists became firmly convinced that exact laws underlying the phenomena of nature actually exist.

Similarly, since the discovery of the chemical structure of benzene by the German scientist Kekule in the nineteenth century, the study of organic structural chemistry has received greater and greater attention. Gradually many organic chemists became interested in the chemistry of living bodies and the different chemical reactions inside living cells. This branch of chemistry became known as biochemistry. More recently, molecular biology has developed as a specialized section of biochemistry.

If nature has laws governing matter, then it is quite conceivable that there must also be laws governing life. The tendency, however, has been to assume that the laws governing life are nothing more than the laws discovered for inanimate matter. Even though the biochemists have discovered chemical processes of ever increasing complexity and apparent sophistication within the living cells, the prevailing assumption has been that all phenomena of life can be accounted for by the same ordering principles that were discovered in the study of simple arrangements of gross matter.

Towards the end of the nineteenth century and the beginning of the twentieth, with the discovery of the fundamental particles—electrons, protons, etc.—the quantum mechanical equations were developed because the earlier equations of classical physics could not describe the behavior of these finer particles of matter. Chemistry and physics have since become more and more unified in the study of atoms and subatomic particles. For example, the earlier concepts of the valence bond theory of chemistry have been dealt with by the atomic and molecular orbital theories derived from quantum mechanics, giving better results in many cases. The well-known Woodward-Hoffmann rule of electrocyclic reactions is also an attempt in this direction.[9] In the present scientific community, a great many scientists are hoping that quantum mechanics will provide the necessary framework for the ultimate understanding of all the phenomena of nature. A summary of the quantum mechanical laws for chemistry is given in Figure 2.

The basic tenet that matter is measurable, calculable, and understandable in terms of physics and chemistry continues to guide the majority of scientists. This reductionistic approach of the physicists and chemists has been borrowed by the molecular biologists and molecular evolutionists, who faithfully assume that life can also be fully understood in terms of atoms and molecules

$$(a) \quad H\Psi = i\hbar \frac{\partial}{\partial t}\Psi$$

$$(b) \quad H =$$

$$\sum_n \frac{-\hbar^2 c^2 \frac{\partial^2}{\partial q_n^2} + \eta_n^2 q_n^2}{2} + \sum_k \frac{-\hbar^2 \nabla_k^2}{2m_k}$$

$$+ \sum_k \frac{i\hbar e_k}{m_k c} \overline{A}(\overline{Q}_k)\cdot\nabla_k + \sum_k \frac{e_k^2}{m_k c^2} \left|\overline{A}(\overline{Q}_k)\right|^2$$

$$- \sum_k \frac{e_k}{2m_k c} \overline{\sigma}_k \cdot \nabla_k \times \overline{A}(\overline{Q}_k) + \sum_{i>j} \frac{e_i e_j - G m_i m_j}{\left|\overline{Q}_i - \overline{Q}_j\right|}$$

$$\overline{A} = \sum q_n \overline{A}_n$$

Figure 2. The laws of nature underlying chemistry.

7

and their interactions. Thus, the well-known physicists Erwin Schrödinger[10] and Niels Bohr[11-12] expressed great hope that life could be completely understood in terms of physics. Similarly, molecular theorists of life such as Watson,[13] Crick,[14] and many others are absolutely convinced that life is a product of chemical reactions. Yet, an analysis of the equations listed in Figure 2 strongly suggests that this reductionistic position is not justified.

We will not discuss these equations in detail here. Readers interested in more details about these equations are referred to Monograph 3 of this series.[15] The first equation is the basic equation of motion, and is known as the Schrödinger equation. This is a second order partial differential equation. The second equation defines the Hamiltonian operator, or the sum total of the various kinetic and potential energy terms believed to be significant in chemical phenomena.[16] According to the viewpoint of the molecular biologists, the final cause underlying all the phenomena of life is represented by these mathematical equations. These equations may thus be taken as expressing the modern scientific view of the absolute truth, at least as far as life is concerned.

Yet if these equations are analyzed conceptually, they are seen to involve nothing more than some simple pushes and pulls between particles. There is a "free radiation" term, which can be visualized in terms of vibrating springs; a kinetic energy term; a term for the "push" between a charged particle and an electromagnetic field; a term for the "pull" between field and magnetic moment; and a term for the pushes and pulls between charged particles. The basic idea is illustrated in Figure 3.

Figure 3 sums up the interactions between molecules as seen in both the classical and quantum mechanical theories. In the quantum mechanical picture, the particles are described by "quantum waves," which give only a statistical estimate of their positions and momenta. It may even be said that in quantum mechanics the very idea of definite particles has become untenable. Nonetheless, the forces governing nature remain the same: just simple pushes and pulls. In summary, then, the modern scientific viewpoint reduces the absolute truth to nothing more than these pushes and pulls between particles: the universe consists of a vast number of particles interacting with one another by simple mechanical rules, having started from some chaotic arrangement. One might well wonder whether mere pushes and pulls can be solely responsible for all the diverse aspects of the world and ourselves that we experience in life. Are molecular biologists and molecular

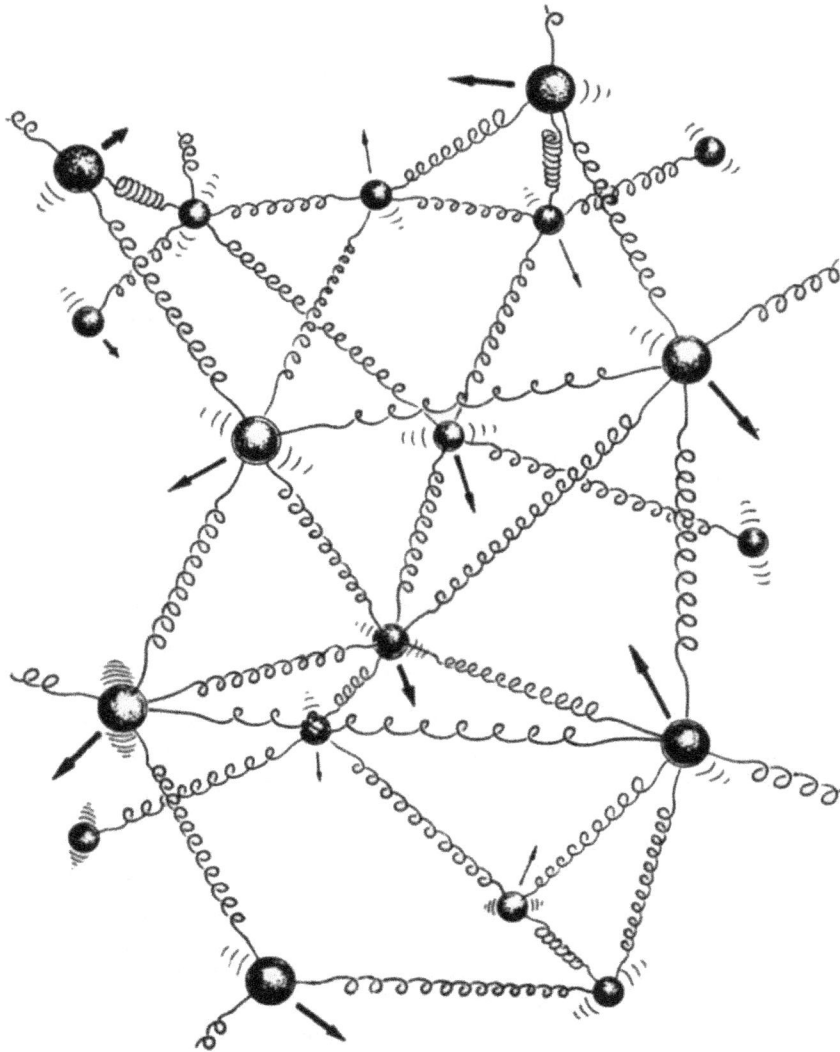

Figure 3. The essence of the laws of nature in modern science.

evolutionists going too far in claiming that life is nothing but a coordinated chemical reaction? What are the motivations and justifications for such a claim? The answers to these questions will be briefly discussed in the following sections.

9

Chemistry and Life: Is the DNA Molecule Life?

Until the nineteenth century, chemists thought that whatever happened in a living system could not be reproduced in the laboratory. In other words, inorganic matter was thought to be fundamentally different from the organic matter composing living material bodies. The prevailing view was that a non-physical vital energy was operating in the living system. In 1828, however, the German chemist Friedrich Wöhler announced the laboratory synthesis of urea from ammonium cynate, an inorganic compound. Urea is an end organic byproduct forming the major solid component of mammalian urine. Wöhler's synthesis of urea profoundly influenced the minds of chemists toward adopting a materialistic view of life. By the late 1850's, Pierre Berthelot reported the production of such organic compounds as alcohols, acetylene, methane and benzene from inorganic chemicals. Gradually, chemists began to think that there was nothing unusual about the organic world. Thus the way was paved for the view that life itself must be made up of chemicals. The announcement of Darwin's theory of evolution in 1859 lent further support to the concept that life was of material origin, a concept that has since remained dominant in both science and philosophy.

About one and a half centuries have passed since Wöhler's synthesis of urea, and indeed organic chemistry has advanced tremendously since that time. Synthetic fibers, synthetic rubber, synthetic dyes, chemotherapeutic agents, synthetic pesticides, synthetic glass and synthetic liquid crystals are some of the major products of synthetic organic chemistry.

Similarly, during the last fifty years or so, many advances have been made in the fields of cell biology and molecular biology. Chemists and biochemists have identified many chemicals such as lipids, proteins, deoxyribonucleic acid (DNA), ribonucleic acid (RNA), hormones and coenzymes inside the cells constituting living material bodies. Many scientists believe that the DNA molecule holds the ultimate explanation of life. It is their genuine hope that once this DNA molecule, the so-called master molecule, is assembled step by step from its constituent atomic elements—carbon (C), hydrogen (H), nitrogen (N), oxygen (O), and phosphorus (P)—their goal of synthesizing life in the test tube will be achieved. This will finally prove that life is, after all, nothing but a system of chemicals. But is the DNA molecule really the essence of life?

We would like to argue that no matter how complex they may be, all molecules or collections of molecules, including DNA, are *dead matter*. What scientists know and agree upon is that the majority of the molecules playing vital roles in living systems are extremely complex. This much is correct. We question only their further conclusion that if complex molecules can somehow be made from simple molecules (for example, proteins from amino acids, and DNA from nucleotides) then life will arise from these complex molecules by virtue of their proper combinations.

Let us briefly examine the chemistry of the cellular DNA molecule. It consists of two intertwined strands of complementary structures forming a regular double helix.[17] From X-ray crystallographic studies, the diameter of the helix is found to be approximately 20 Å, and each strand makes a complete turn every 34 Å (or every 10 nucleotides).[18] Strings made of alternate groups of phosphate and sugar (deoxyribose) form the backbone of the two strands. Each phosphate group links to deoxyribose, a five-carbon chain sugar. The sugar in turn links to one of two possible bases of purine (guanine or adenine) or two possible bases of pyrimidine (thymine or cytosine) through hydrogen bonds. Adenine (A) is always paired with thymine (T), and guanine (G) with cytosine (C), for conformational reasons and because of the donor-acceptor natures of the hydrogen bonding groups. As a stereochemical consequence of this strict base pairing, the two polynucleotide chains run in opposite directions. These are illustrated in Figure 4. Although hydrogen bonding between other base pairs is possible, it leads to nucleotide pairs which have the wrong external geometry and do not fit into the regular double-helical structure.

This strict requirement of base pairing is responsible for the systematic replication process of DNA. Geneticists commonly assume that DNA is the carrier of the genetic information of the cell. It duplicates itself before cell division to provide each daughter cell with a complete set of DNA molecules. DNA replication involves strand separation, and each separated strand forms the template for the condensation of a complementary strand. This is commonly called the Watson-Crick mechanism.

Descriptions such as this of DNA and its replication mechanism are commonly given as though they provided a complete description of the most fundamental processes of life—a final mechanical, step-by-step breakdown of these life processes into understandable chemical terms. However, this is far from true. An enormous gulf

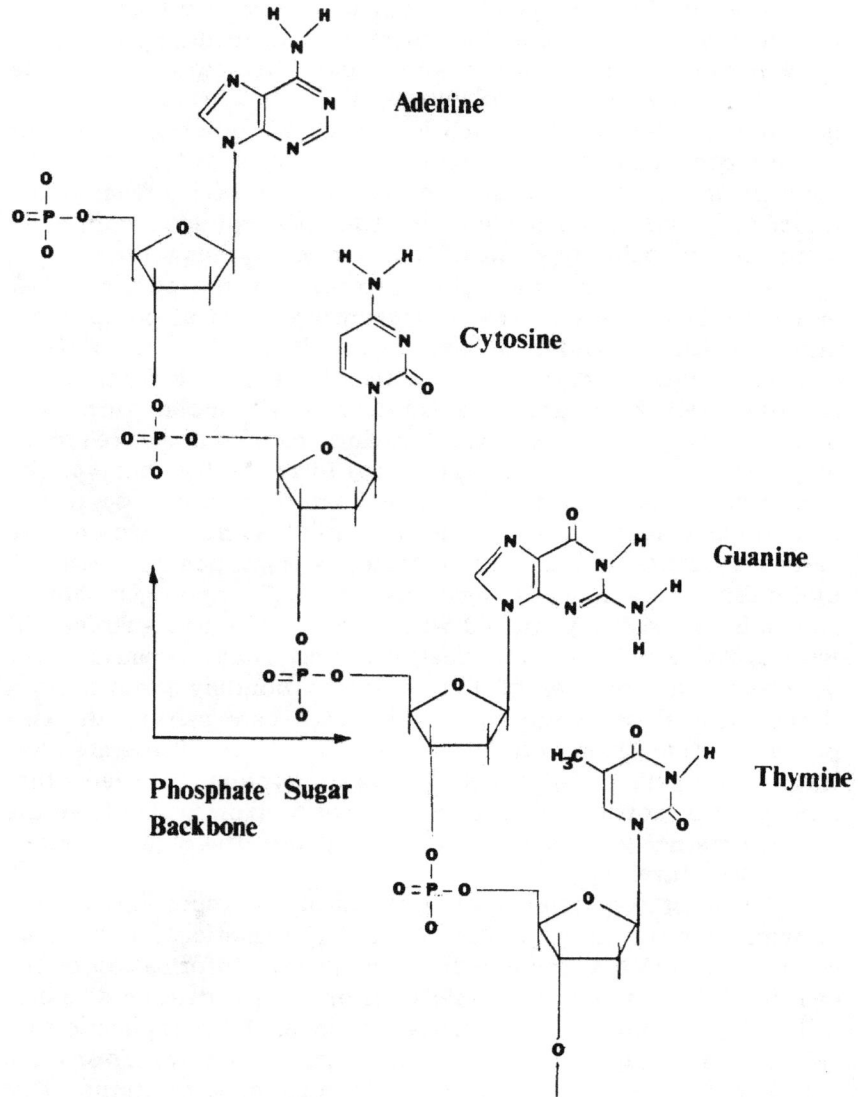

Figure 4. A section of the DNA molecule.

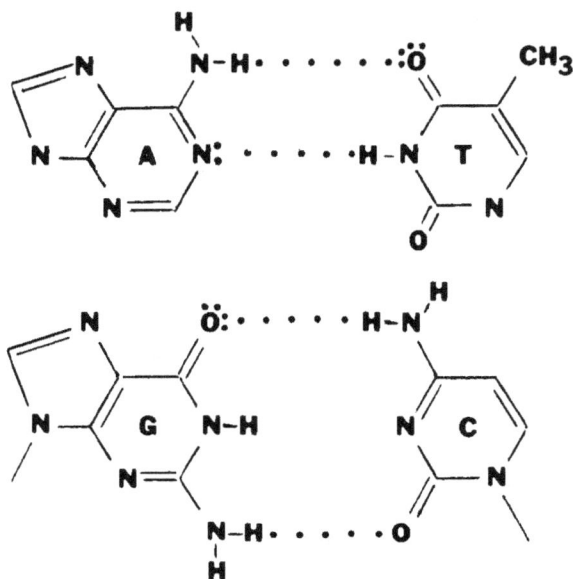

Figure 5. Hydrogen bonded base pairs in the DNA molecule.

lies between the few simple chemical facts known about DNA and the actual functioning of a cell. All that science actually knows about DNA are a few relations between inanimate chemicals. The gap between this knowledge and an actual chemical understanding of life is bridged only by faith.

To illustrate this, let us consider a few features of the cellular reproduction process. According to Watson, the replication and maintenance of DNA requires at least four different kinds of enzymes: endonuclease, exonuclease, DNA polymerase, and polynucleotide ligase.[19] These are all practically unknown at present. The one which seems to be best understood, DNA polymerase, is estimated to contain some 1,100 amino acid subunits, but the arrangement of these subunits is still unknown. It is thought that DNA polymerase is involved in the cellular replication of DNA. However, there is also evidence that this enzyme is involved in the repair of damaged DNA instead, and that in bacteria other, unknown enzymes in the cell membrane are required for replication.[20]

The replication process is thus very poorly understood. For example, the single chromosome of *E. coli* is thought to be a loop of double-stranded DNA some 500 times longer than the cell itself. Due to its spiral nature, this tangled loop must spin on its axis some 360,000 times in the course of a single replication, and the two loops must be neatly separated. In order to account for this, biochemists have postulated many different molecular mechanisms, but none of these are clearly understood.[21] This is just one of many examples that we could cite.

Although we may imagine that the cell is nothing more than an elaborate chemical machine, we actually do not at all know how this machine works. We have no idea how the large scale actions of a cell (what to speak of a multicellular organism) can be reduced to the reactions of molecules. Indeed, we do not even fully understand the chemical interactions of water molecules, and the operations of enzymes composed of 1,100 amino acids are certainly a mystery.

The assumption that the cell is a machine running according to simple push-pull laws is, therefore, simply a matter of faith. It may be imagined that thousands of reactions of the form $A_i + B_i \rightarrow C_i$ can combine to create an elaborate chemical automaton surpassing even the most sophisticated man-made computers. However, in contemplating this analogy we should consider that even the most detailed knowledge of the intricate functioning of a computer would be incomplete unless it entailed an understanding of the programmer. In like manner, it is quite possible in the context of current knowledge that other laws are involved in the operation of cells that are unknown to modern chemistry. The most that can be said at present is that the knowledge of the biochemists is a knowledge of chemical reactions; it cannot be claimed that it constitutes an understanding of life.

As a further example, consider the recently reported synthesis of the *E. coli* gene that codes for tyrosine transfer ribonucleic acid (tRNA).[22] This gene has only 126 nucleotides, and commercially synthesized nucleotides are used as the starting materials for the gene's synthesis. The nucleotides are chemically hooked to form di-, tri-, and tetranucleotides. These units are further chemically assembled into dioxyribooligonucleotide segments of 10 to 15 units. The segments that possess complementary base sequences are enzymatically connected with DNA ligase to form larger duplexes, which are themselves finally connected enzymatically to complete the synthesis. (It is not a total chemical synthesis, in the sense that the natural enzyme has to be used to join the larger units.)

A gene is taken as a fundamental unit of heridity. According to geneticists, everything from the color of rose petals to the shape and color of human eyes is determined by genes. It has been reported that the functioning of this artificial gene could be detected in a living cell. These are quite significant findings so far as chemical knowledge is concerned. They suggest the possibility that a geneticist will be able to manipulate genes chemically, replacing defective ones with healthy ones. This does not, however, demonstrate that genes are completely responsible for life. Rather, it simply indicates that cells make use of messages coded in chemical form, and that our technology may enable us to take advantage of this medium of information storage.

At this stage of scientific knowledge, all the experimental techniques and tools needed to synthesize most of the chemicals primarily found in living cells (for example, proteins, hormones, lipids, carbohydrates, vitamins and genes) are available. Yet scientists are nowhere near to constructing a complete "synthetic living cell" in the test tube. The great hope expressed by many molecular biologists about a quarter century ago (after the historic discovery of the double helical structure of DNA by Watson and Crick) seems to have faded away in the midst of new discoveries.

Indeed, the findings of the biochemists, far from proving that life is a chemical phenomenon, have strongly demonstrated that the present scientific understanding of life is inadequate. In Darwin's time the cell appeared to be little more than a simple bag of organic compounds that one might readily hope to describe in chemical terms. The enormous complexity encountered in recent biochemical investigations, however, has shown that this hope is unrealistic. Modern science is far from having understood the principles of life.

Szent-Györgyi, the Nobel-prize-winning chemist, thus remarked: "in my search for the secret of life, I ended up with atoms and electrons which have no life at all. Somewhere along the line, life has run out through my fingers. So, in my old age, I am now retracing my steps . . ."[23] This is the basic point we would like to emphasize. Atoms and molecules are lifeless. A gene or a DNA molecule is not life, and a protein molecule is not life. Indeed, we propose that a collection of these molecules is also not life. The recent announcement of Khorana's synthetic gene is not different from that of Wöhler's synthesis of urea in 1828 as far as our understanding of life is concerned.

Chemical Evolution: A Molecular Fairy Tale?

The theory of chemical evolution rests upon three assumptions: (1) The hypothetical primitive atmosphere must have been either reducing or neutral. This means that there was no free oxygen in the atmosphere in the earth's distant past. (2) Simple molecules like amino acids, purines, pyrimidines, and sugars were formed within this atmosphere under the action of ultraviolet radiation, electrical discharges, radioactivity, thermal energy, and so on. (3) In the course of time these molecules gave rise to protoproteins, protonucleic acids, and other protocellular components, which in turn gave rise to the so-called protocells and finally to the living cell.

We can briefly analyze these assumptions by purely scientific reasoning and argument. It is a foregone conclusion of many molecular evolutionists that the primitive atmosphere consisted of carbon (C) in the form of hydrocarbon such as methane (CH_4), nitrogen (N) in the form of ammonia (NH_3), oxygen (O) in the form of water (H_2O), and sulfur (S) in the form of hydrogen sulfide (H_2S). This was first proposed by Oparin,[24] the Russian evolutionist, and Urey,[25] the American physicist.

Based on this assumption, Miller[26] performed an experiment in which he passed an electric discharge through a gaseous mixture of methane, ammonia, hydrogen, and water vapor. Amino acids such as glycine, alanine, aspartic acid, and glutamic acid were observed as some of the components of the reaction products. Since amino acids are the smallest units of the protein molecule, Miller's experiment gave the molecular evolutionists great hope and encouragement for their idea of the chemical origin of life. They claim that such steps are the ones that will finally lead to life. However, to accept this claim as proven would be quite premature.

The idea of the primitive reducing atmosphere has received strong and serious criticisms from scientists of various disciplines. Their arguments suggest overwhelming drawbacks in the conjecture. Available data from geology, geophysics and geochemistry argue strongly against this idea. Abelson,[27] for example, argues that there is no evidence for the reducing atmosphere, and that ammonia would have quickly disappeared because the effective threshold for degradation by ultraviolet radiation is 2,250 Å. He suggests that a quantity of ammonia equivalent to the present atmospheric nitrogen would be destroyed in approximately 30,000 years.

16

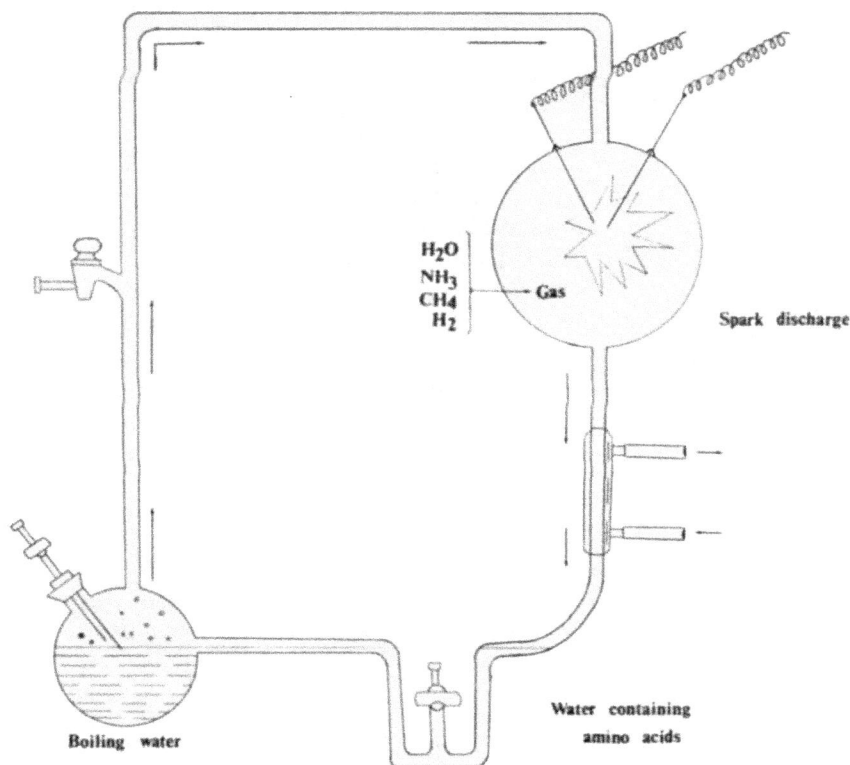

Figure 6. Miller's experiment.

Abelson has also suggested that if the primitive atmosphere contained large amounts of methane gas, geologic evidence for it should be available. Laboratory experiments show that irradiating a highly reducing atmosphere produces hydrophobic organic molecules that are absorbed by sedimentary clays. This suggests that the earliest rocks should have contained an unusually large proportion of carbon or organic chemicals. However, this is not the case.

From observations based on the stratigraphical record, Davidson[28] concludes that there is no evidence that a primeval reducing atmosphere might have persisted during much of Precambrian time. Brinkmann[29] shows from theoretical calculation that dissociation of water vapor by ultraviolet light must have generated enough oxygen very early in the history of the earth to create an oxidizing atmosphere.

17

In light of these arguments, the idea of a primeval reducing atmosphere does not seem tenable. Of course, this does not mean the end of speculation on the chemical origin of life. Although the reducing atmosphere has been by far the most popular, many other hypothetical primitive atmospheres have been proposed.

Thus, the gaseous mixture in Miller's experiment can be replaced by a mixture of carbon monoxide, nitrogen, hydrogen, and water vapor, giving comparable results and thus indicating that the carbon need not be in the form of hydrocarbon gas.[30] The molecular evolutionist Matthews[31] has advanced another theory about the possible formation of protein from hydrocyanic acid (HCN) gas. Electrical discharge experiments in a mixture of nitrogen, carbon monoxide, and hydrogen give HCN as one of the principal products.[27] HCN is an even more promising candidate as far as the formation of proteins, purines, pyrimidines, and other molecules of biological importance is concerned.

One can arrive at many alternative theories about the unknown past, and these can be criticized in turn. (For example, the two atmospheres just mentioned would not endure if the dissociation of water vapor generated substantial amounts of free oxygen.) But, where is the truth? We can only conclude that conditions (1) and (2) are shaky and speculative assumptions at best.

It has been claimed that the so-called coacervates of Oparin[32] and the proteinoid microspheres of Fox[33] are the protocells. We would like to examine what these words mean chemically. By definition, a coacervate is an aggregate of colloidal droplets held together by electrostatic charges. Coacervate formation has been observed when large molecules possessing hydrophobic and hydrophilic sites are dissolved in water. They consist of spheres or droplets separated from the bulk solution. It is believed that coacervates are the end product of the reduction of the hydration layer surrounding colloidal particles.

The phenomena of coacervate formation were first studied in detail by Bungenberg de Jong,[34] who demonstrated that coaceration is an effective technique for concentrating compounds of high molecular weight from aqueous solutions. The coacervate droplets are usually obtained by mixing solutions of proteins and other polymers—for example, solutions of gelatine and gum arabic, solutions of various proteins and nucleic acids, and so on. Oparin has reported that in the synthesis of polyadenine in vitro in a polypeptide solution, coacervate droplets begin to separate from the bulk solution as soon as the molecules reach a certain size.[35] He further draws the conclusion that non-specific polymerization of organic compounds must have taken place in the "primeval broth," leading to the formation of polypeptides and polynucleotides with randomly arranged monomeric residues in their chains. These polymers might have separated in the form of coacervate droplets, thus creating isolated systems where further evolution of organic polymers might have occurred that was not possible in the solution as a whole.

Oparin suggests that as soon as the polynucleotide chain reaches a certain size, even though it has a disorderly structure, it will interact with polypeptides and other compounds in the "primeval nutrient broth" and separate out from the solution in the form of coacervates. His reasoning is that although there could not be any selection of individual nucleotide molecules when they were in simple aqueous solution, the situation is different when they separate out as coacervate droplets after interacting with polypeptides. Because of the double helical character of the two complementary chains of polynucleotides, their inclusion in coacervate droplets (or protobionts) may have had certain effects on the polymerization of the amino acids in those systems. Those arrangements of amino acids unfavorable for the increasing catalytic activity of the polypeptides would be destroyed by natural selection. In this way, the structure of the protein-like polypeptides, and also that of the polynucleotides controlling their synthesis, may gradually have become more ordered and better adapted.

This may sound convincing until we look at the scientific facts. Coacervate formation is similar to the well-known chemical process called "salting out." For example, if the salt potassium chloride is added to a soap solution of potassium oleate, the phenomenon of coacervate formation is exhibited. The hydrocarbon chain of this soap molecule is less soluble in water. If increasing amounts of potassium chloride are added to a concentrated soap solution, two layers (phases) will form, and just before the separation of these distinct layers, oily droplets will appear. These are termed coacervates. The explanation is that the potassium chloride molecules compete with the water molecules in the potassium oleate solution, thus allowing the water molecules to separate from the hydrophobic chain of the oleate moiety. In chemical language these droplets are commonly known as spherical

19

micelles. In aqueous solution, the nonpolar (hydrophobic) portion of the monomers reduce their contact with water and form the micellar core, while the polar (hydrophilic) portions remain in contact with water, forming roughly spherical micelles. In some nonaqueous (nonpolar) solvents the reverse phenomenon is observed. The polar groups of the monomers may become solvophobic, thereby forming the cores of the micelles. These are called inverted micelles. Cylindrical or lamellar aggregates also result in highly concentrated solutions. The two types of micelles are illustrated in Figure 7.

Monomers and micelles are usually in rapid dynamic equilibrium, and micelles are known to catalyze chemical reactions. Thus one can safely conclude that Oparin's coacervates simply exhibit the phenomena characteristic of micellar chemistry. Apart from his many "may have beens," he is simply describing a few physical properties of inanimate matter. Fox, Oparin's own colleague, has criticized his conjectures about these coacervates: "besides failing to answer the crucial primordial question, they are neither uniform nor stable."[36]

Fox, on the other hand, claims that his so-called protenoid is the "molecular missing link between prelife and life."[37] But, we can also show that this claim is completely erroneous and unfounded.

Proteinoids are formed by pyrocondensing dry amino acids.

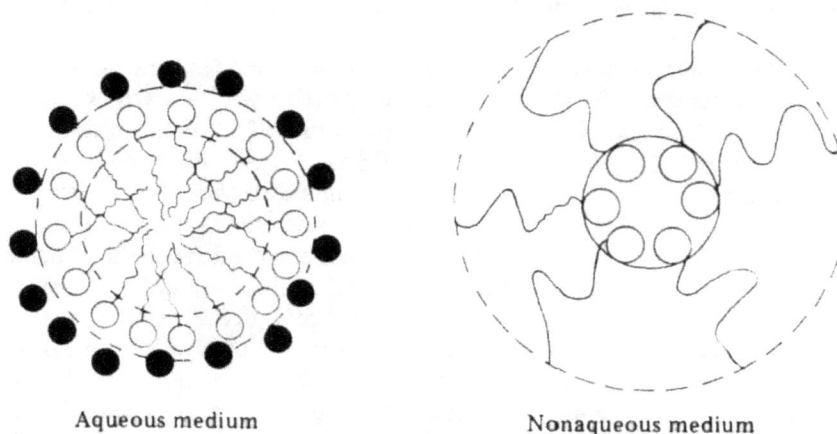

Aqueous medium   Nonaqueous medium

Figure 7. Spherical micelles.

The amino acids are heated at 140° to 180° C for about six hours. Only a sufficient proportion of one of the acidic amino acids, aspartic and glutamic acids, or of the basic amino acid, lysine, is required. The reaction mixture is treated with hot water and, after being stirred, the water-insoluble material is separated by filtration. When the filtrate cools down, the product precipitates as microscopic globules that Fox calls proteinoid microspheres. The molecular weights of the products can be as high as 20,000 when glutamic acid is heated with glycine. The proteinoids give all the color test results common to proteins. Fox further claims that the amino acid sequences in these proteinoids are highly nonrandom. Proteinoids catalyze like enzymes in hydrolysis of esters, decarboxylation, amination, and deamination reactions. He also claims that these proteinoids multiply by division in a manner similar to that of living cells.

We would like to suggest that all the above properties are simply the physico-chemical properties inherent in such molecules. They have nothing to do with the characteristics of living cells. Chemically, it is expected that when a mixture of amino acids is heated at elevated temperatures, polymers will be formed. These are the peptides, and they show the properties inherent in proteins. However, Fox's argument for the nonrandom sequencing of the amino acids in his reaction is quite objectionable. As a matter of fact, some of his own supporters accuse him of deception. Miller and Orgel in this respect remark:

> ... the degree of nonrandomness in thermal polypeptides so far demonstrated is minute compared with the nonrandomness of proteins. It is deceptive, then, to suggest that thermal polypeptides are similar to proteins in their nonrandomness.[38]

They continue by saying:

> The importance of these thermal syntheses in prebiotic chemistry is a very controversial matter. We do not believe that they were very important because we doubt that polypeptides could have been synthesized in large quantities at the surface of the earth by thermal reactions of the kind so far demonstrated.[39]

So many unique events and conditions have to be simultaneously fulfilled in Fox's model of proteinoid formation that it is very doubtful whether many chemists will ever take it seriously.

21

First of all, the temperature specified by Fox for the heating of the amino acids is very unlikely to occur on the surface of the earth. Although the temperature in some hot springs may rise to 140° or 180° C, such reactions are extremely improbable. Fox's conditions require that the amino acids be in the right place and also be dry. The polymerization reaction of amino acids does not take place in the presence of water. In fact, the reverse reaction will be favored, and the polypeptides will be completely hydrolysed to amino acids under such conditions. The thermodynamic free energy of this condensation reaction is about 2.00 to 5.00 kcal/mole, which means that the reaction is very unfavorable towards the product side.

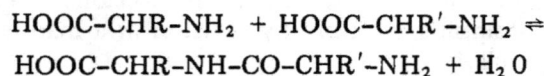

$$HOOC-CHR-NH_2 + HOOC-CHR'-NH_2 \rightleftharpoons$$
$$HOOC-CHR-NH-CO-CHR'-NH_2 + H_2O$$

$$\Delta G = 2\text{-}5 \text{ kcal/mole}$$

The other possibility of temperatures as high as 140° to 180° C is in volcanoes. Here again the conditions are not favorable for the production of the polypeptides. In volcanoes the temperature of molten lava is about 1,200° C, which will completely destroy the amino acids. It should also be mentioned that ultraviolet radiation, being a very powerful source of energy, can not only create organic molecules but also destroy them—especially macromolecules such as proteins and nucleotides.

Finally, from a purely chemical point of view, Fox's proteinoids may be expected to possess some catalytic activity as general acid-base catalysts according to a catalysis law of the Brönsted type. This has nothing to do with the true nature of a living cell and, therefore, with life. One can see that all these claims amount to no more than *molecular fairy tales*. They are like taking a rope to be a serpent. Thus, from the above evidence, we can see that assumptions (2) and (3) have no valid scientific foundation.

Chemical Evolution:
The Role of Chance and the Long Time Span

It is often claimed in books on the origin of life that life is bound to evolve from chemicals, given sufficient time. The idea is that even though an event may be highly improbable, if one waits long enough it is bound to happen. As one popular account has it,

> The odds against the right molecules being in the right place at the right time are staggering. Yet, as science measures it, so is the time scale on which nature works. Indeed, what seems an impossible occurrence at any one moment would, given untold eons, become a certainty.[39]

Therefore, we shall briefly examine how chance applies to the formation of biological macromolecules.

Let us consider the probability of forming a particular chain of amino acids at random. Since there are 20 kinds of amino acids in living bodies, there are some 20^N possible proteins composed of N amino acids linked in a chain. Of these, there are

$$\binom{N}{N/10} 20^{N/10}$$

chains equal to a particular given chain for all but 10% of its amino acid links. If we pick an N link chain completely at random, the probability that 90% of it will match our given chain is therefore,

$$P = \binom{N}{N/10} 20^{-9N/10} \approx \frac{10^{-N}}{\sqrt{N/2}} \tag{1}$$

Figure 8 illustrates these probabilities for a number of proteins found in living cells. These probabilities show that it is extremely unlikely that a particular protein will form by chance, even if we allow for 10% of its amino acid subunits to be in error. These conclusions are elementary and well-known. However, numbers such as 10^{-52} and 10^{-246} are not very meaningful by themselves. In order to see what these numbers mean, let us consider them in the context of a model for the chemical origin of a proto-

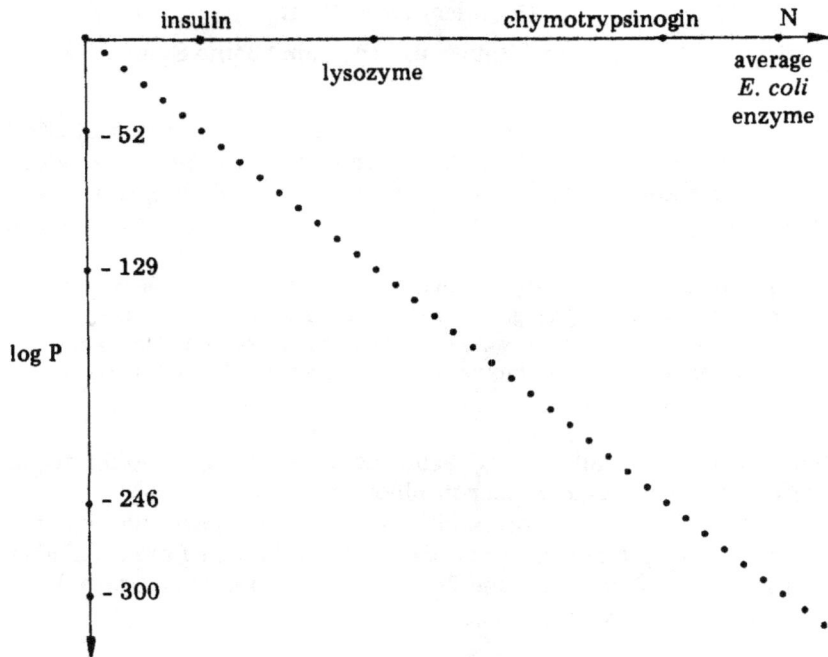

Figure 8. The probability, P, of randomly picking a specific protein of N amino acids with an error of 10% or less.

cell. We would like to show that the necessary molecular components for a functioning protocell cannot be expected to come together by chance, even if thousands of billions of years are allowed.

For our model, let us postulate a "primordial soup" one kilometer thick, covering the entire surface of the earth. Let us suppose that this soup is so packed with protein molecules that there is an average of one protein in each $20 \times 20 \times 25$ Å^3 box throughout its entire volume. We shall assume that these proteins are continuously being created, destroyed, and moved about, so that their arrangement changes in each millionth of a second. We shall also assume that these arrangements are completely random and disorderly. This model is intended to provide more random arrangements of molecules per unit time than could ever have been produced in any actual situation on the primordial earth. As such it takes into account both the random creation and the random diffusion of molecules.

24

We are interested in seeing whether or not the necessary initial constituents of a self-sustaining protocell could be expected to come together by chance in this soup over a long period of time. In order to do this, we shall first review the molecular composition of living cells. The distribution of molecules in a cell of the bacterium *Escherichia coli* is outlined in Figure 9.[41] We can see from this table that many thousands of very large protein molecules are involved in the metabolism of *E. coli*, even though this is one of the smallest independent living cells. The average size of these proteins is about 300 amino acid subunits.

An example of the function of such complex proteins in cells is the process of biosynthesis of L-isoleucine illustrated in Figure 10. In this process, L-isoleucine is produced from L-threonine in five steps.[42] Each step is catalyzed by a specific large protein molecule called an enzyme. Such enzymes have the property of greatly accelerating a particular chemical reaction, while not affecting other reactions at all. They also must be capable of interacting with other particular molecules to regulate their activity. In this example, the enzyme L-threonine deaminase, catalyzing the first step in the chain, is sensitive to the presence of the product molecule L-isoleucine, produced four steps later. When the concentration of L-isoleucine reaches a certain critical level, this enzyme ceases to function, insuring that no more than the necessary amount of the product is formed. We can thus see that a precise and integrated system of molecular interactions analogous

Cell Component	Approximate Number/Cell	Different Kinds
Water	4×10^{10}	1
Inorganic ions	2.5×10^8	20
Carbohydrates and precursors	2×10^8	200
Amino acids and precursors	3×10^7	100
Lipids and precursors	2.5×10^7	50
Nucleotides and precursors	1.2×10^7	200
Proteins	10^6	2000 to 3000
DNA	4	1
tRNA	4×10^5	40
mRNA	10^3	1000

Figure 9. The distribution of molecules in *Escherichia coli*.

$$\underset{\underset{H\ \ \ \ H}{|\ \ \ \ |}}{CH_3-\overset{\overset{OH\ \ NH_3}{|\ \ \ \ |}}{C}-C-COO}\qquad\qquad\text{L-Threonine}$$

L-Threonine deaminase

$$CH_3-CH_2-\overset{\overset{O}{||}}{C}-COO^-\qquad\qquad\propto\text{-Ketobutyrate}$$

$$CH_3-\underset{H}{\overset{|}{C}}=O$$

$$CH_3-CH_2-\underset{\underset{CH_3}{|}}{\overset{\overset{OH}{|}}{\underset{C=O}{C}}}-COO^-\qquad\qquad\begin{array}{l}\propto\text{-Aceto-}\propto\text{-hydroxy-}\\ \text{butyrate}\end{array}$$

NADH

NAD$^+$

$$CH_3-CH_2-\underset{\underset{OH\ OH}{|\ \ \ \ |}}{\overset{\overset{CH_3\,H}{|\ \ \ \ |}}{C-C}}-COO^-\qquad\qquad\begin{array}{l}\propto,\beta\text{-Dihydroxy-}\\ \beta\text{-methylvalerate}\end{array}$$

H$_2$O

$$CH_3-CH_2-\underset{\underset{H\ \ \ \ O}{|\ \ \ \ ||}}{\overset{\overset{CH_3}{|}}{C-C}}-COO^-\qquad\qquad\propto\text{-Keto-}\beta\text{-methylvalerate}$$

Transaminase

$$CH_3-CH_2-\underset{\underset{H\ \ \ \ H}{|\ \ \ \ |}}{\overset{\overset{CH_3\,NH_3}{|\ \ \ \ |}}{C-C}}-COO^-\qquad\qquad\text{L-Isoleucine}$$

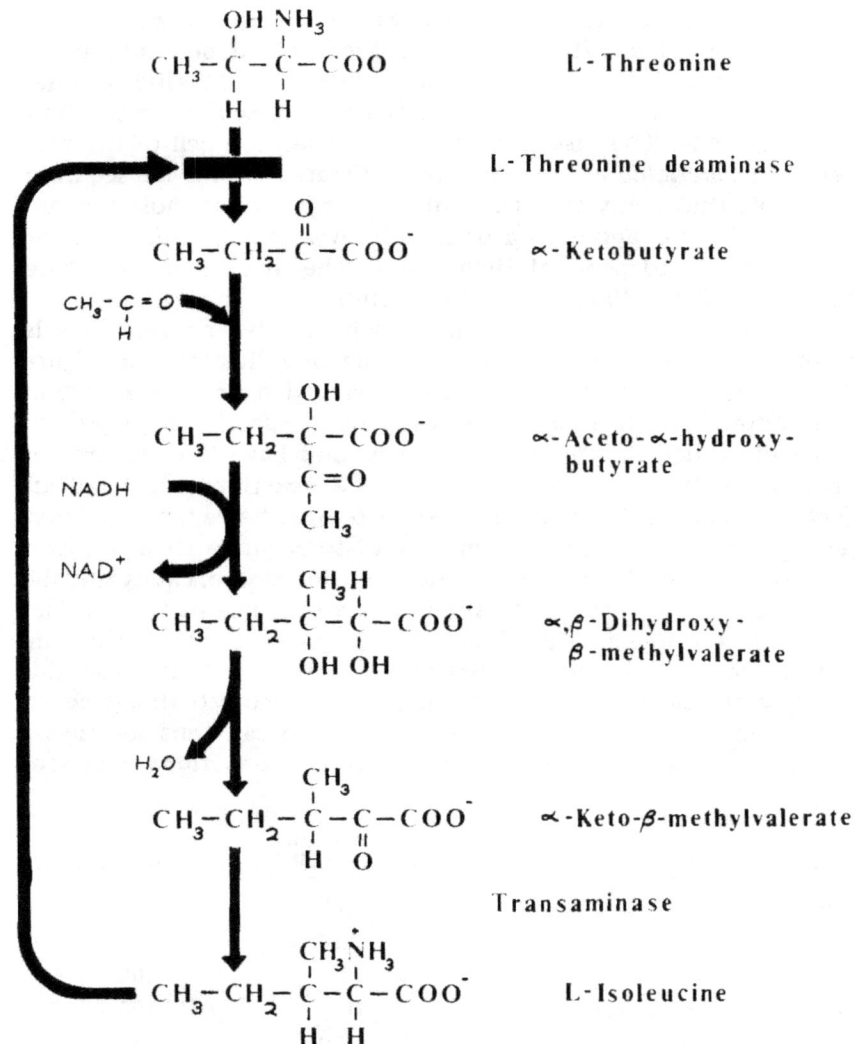

Figure 10. Enzymatic regulation in cellular synthesis.

to a sophisticated computer program is necessary for the harmonious functioning of a self-sustaining metabolic system.

Another example is the system of enzymes involved in the replication of DNA. Here we are indeed faced with an intriguing problem. In order for DNA to replicate, certain highly complex enzymes are needed which are themselves encoded in DNA.[43]

How could such an arrangement get started? These are both examples in which many interdependent cellular components interact in such a way that the successful functioning of the whole cannot take place unless all of the components are present.

We do not know whether any kind of self-sustaining metabolizing unit is possible with substantially fewer and smaller molecules than a small bacterial cell like *E. coli.* Yet, for the sake of argument, let us consider a protocell made with protein molecules no larger than insulin. We choose insulin for the sake of clarity. Its structure has been analyzed, and with only 51 amino acid subunits, it is one of the smallest known proteins. We should note that the arrangement of the amino acids in insulin has to be very precise in order for it to function properly in the human body. The British scientist Sanger demonstrated that a very slight modification of any one of its molecules—for example, the removal or modification of any one of the amino acids—would spoil its activity as an anti-diabetic agent.

Even though a protocell might be less efficient than modern organisms, it would have to have a sufficiently sophisticated metabolic apparatus to enable it to continue functioning for a long period of time—long enough to give it the opportunity to evolve further by the accumulation of mutations and adaptations. Since many highly specific protein molecules are evidently needed in the functioning of present-day living cells, it stands to reason that similar molecules should be required in our hypothetical protocell.

Let us compute the probability that k specific proteins of N amino acid units apiece will appear in a small volume in our "primordial soup" at some time in 1,000 billion years. We shall take this volume to be about 10^{12} Å^3, the approximate volume of an *E. coli* cell. Neglecting insignificant terms, this probability is no more than

$$P_r \leqslant 1.6 \times 10^{61} \ (10^{10} \ P)^k/k! \qquad (2)$$

where P is the probability for the random formation of one protein of N amino acids.[44] If we use equation (1) for P and let N = 51, the number of amino acids in insulin, we find that

$$P_r \leqslant 10^{-21} \qquad (3)$$

even for k = 2. This means that two specific molecules the size of insulin could not be expected to occupy by chance the same volume of *E. coli* size anywhere in our super-concentrated soup at any time in billions of billions of years.

Even if we let N = 20, we find the similar probability of finding six specific peptides of 20 amino acids apiece in one volume the size of *E. coli* to be

$$P_r \leqslant .02 \qquad\qquad (4)$$

This probability indicates that there is only one chance in 50 for these six peptides to be found together even once in a thousand billion years.

These figures, and others which can be easily calculated, show that chance cannot be expected to bring together the initial components of a protocell, even though thousands of billions of years are allowed for this process. We should point out that these calculations are by no means limited to proteins. They also apply to the formation of the specifically ordered nucleotide chains that are believed to have been essential components of the protocells.

It is ironic that even though these calculations are both elementary and familiar to many scientists, students are nonetheless taught in schools and colleges that life has arisen by a chance combination of molecules over a long span of time. Evidently, however, chance will not suffice; some other cause must be invoked to account for the structures of living cells. The only alternative available within the limits of modern science is to suppose that the simple push-pull laws of molecular interaction can somehow bring cells together out of chaos. But how are they to do this? No explanation ever proposed has stood up under scrutiny. Yet the teaching that life has arisen from matter continues without reservations.

Chemical Evolution: Intellectual Dishonesty

Many scientists of the materialistic evolutionary persuasion seem to lack a genuine spirit of intellectual honesty. But unfortunately their theory is on the rise and their influence is strong. They are widely respected as authorities, although they are spreading false knowledge in the guise of science. Such dangerous dishonesty cripples the spirit of true knowledge and misleads young intellects.

Many of the evolutionary scientists try to push forward their own theories, right or wrong. For example, even though Fox's theory of microspheres has been extensively criticized by his own colleagues as unsound, it has nonetheless been widely advertised as a demonstration of the origin of life. In introductory texts aimed at college and high school students one can find glowing accounts of this theory devoid of any critical commentary.[45]

The same thing can be said for practically every aspect of the modern scientific account of life. The biochemical accounts of DNA and RNA are often presented in popular literature as though all the significant aspects of their functioning were clearly understood in chemical terms. This literature stems from the professional scientists who work in these fields. Yet, if one peruses their more technical work, as we have briefly done, one finds a labyrinth of difficulties and mysteries. It is plain that the chemical theory of life has not been established, but one must look very far through the literature of modern science to find a frank admission of this.

Thus, although evolution is far from proven, many scientists speak as though it were an established fact. George Wald says:

> Surely this is a great part of our dignity as men, that we can know, and that through us matter can know itself; that beginning with protons and electrons, out of the womb of time and the vastness of space, we can begin to understand; that organized as in us, the hydrogen, the carbon, the nitrogen, the oxygen, those 16 to 21 elements, the water, the sunlight—all, having become us, can begin to understand what they are, and how they came to be.[46]

The conclusion drawn is bold, but the evidence is too weak to be scientific.

There is, unfortunately, evidence that many scientists are at least as interested in using propaganda and political pull to in-

29

doctrinate people as they are in the pursuit of genuine knowledge. It is often stressed that the theories of science stand on the basis of evidence and reason alone, and that these theories are always open to challenge. Yet we have recently seen over 150 prominent scientists signing a petition "affirming evolution as a principle of science," and declaring that, "There are no alternative theories to the principle of evolution . . . that any competent biologist today takes seriously."[47] Statements such as this reflect a marked lack of openness in modern scientific circles to the serious discussion of alternatives to prevailing views.

They also reveal a willingness to use theories belonging in the realm of conjecture and speculation as tools for the manipulation of people's behavior and thinking. Thus, the Nobel-prize-winner Francis Crick states:

> Once one has become adjusted to the idea that we are here because we have evolved from simple chemical compounds by a process of natural selection, it is remarkable how many of the problems of the modern world take on a completely new light. It is for this reason that it is important that science in general, and natural selection in particular should become the basis on which we are to build the new culture.[48]

What will this new culture be? Obviously it will be a molecular culture that will produce men like Aldous Huxley, who said,

> I had motives for not wanting the world to have meaning . . . For myself, as, no doubt, for most of my contemporaries, the philosophy of meaninglessness was simultaneously liberation from a certain political and economic system and liberation from a certain system of morality.[49]

Motives such as these can also be seen in such building blocks of the "new culture" as *The Origin of Johnny*[50] and *Biology Today*.[51] These books, the first intended for young teenages and the second for beginning college students, both present a strictly molecular view of life and conclude by advocating its logical corollary: a philosophy of complete hedonism.

In view of the consequences of such a philosophical choice, it is unfortunate that it should be given such strong support through the uncritical propagation of highly questionable theories among innocent people. We can only conclude that in studying the origin of life, many scientists have failed to act in the true spirit of knowledge. One could appreciate their efforts more if the findings were presented more honestly.

Chemical Evolution: Implications and Challenges

Molecules lack inherent purpose and meaning. Yet we give value to life. A reciprocal feeling of love and care exists among people and among other living entities. Parents think about their children; a nation thinks about the welfare of its subjects. Great sages think about the welfare of all living entities—starting from an ant up to man. There can be no value without purpose and meaning. However, the doctrine of the chemical nature of life reduces life to complete meaninglessness. Since this is contrary to the truth, it generates a sense of emptiness and unhappiness in one's subliminal mind. This is vividly illustrated in the case of Darwin, the father of the doctrine of evolution. He developed, in his own words, a "curious and lamentable loss of the higher aesthetic tastes."[52] He expressed this loss in his autobiography:

> I have said in one respect my mind has changed during the last twenty or thirty years. Up to the age of thirty, or beyond it, poetry of many kinds, such as the works of Milton, Gray, Bryon, Wordsworth, Coleridge, and Shelley, gave me great pleasure, and even as a schoolboy I took intense delight in Shakespeare ... I have also said that formerly pictures gave me considerable, and music very great delight. But now for many years I cannot endure to read a line of poetry: I have tried lately to read Shakespeare and found it so intolerably dull that it nauseated me. I have also almost lost my taste for pictures or music ... My mind seems to have become a kind of machine for grinding general laws out of large collections of facts, but why this should have caused the atrophy of that part of the brain alone, on which the higher tastes depend, I cannot conceive ...... The loss of these tastes is a loss of happiness, and may possibly be injurious to the intellect, and more probably to the moral character, by enfeebling the emotional part of our nature.[53]

It is ironic that Darwin should have expressed these thoughts. Why should man, if he is a product of molecular pushes and pulls, worry about happiness or unhappiness? Why should people busy their minds about moral and ethical values? Why should it be necessary to establish educational institutions? Why do problems like those of disease, drugs, alcoholism, smoking, violence, crime, abortion, and euthanasia bother our minds?

Educated people have shown a renewed concern for professional ethics and human values.[54] This is a direct challenge to the

32

Figure 11. Charles Darwin.

doctrine of the molecular character of life. Great concern has been expressed over: the chlorofluorocarbon controversy—skin cancer may be caused by the depletion of the ozone layer in the stratosphere; the ban of synthetic sweetners—cyclamate and saccharin, for example, may cause cancer; recombinant DNA research—when a gene is transferred from one organism to another, harmful and

33

uncontrollable organisms might escape. The first Bioethics Center was formed at Georgetown University, Washington, D.C., in 1971 for concerned scientists who wanted to study perplexing biomedical problems like genetic engineering and organ transplants. [55] A growing concern has developed over bioethical problems such as *in vitro* fertilization of human eggs and their implantation, cloning, and so on. Because of such concerns, the Illinois Institute of Technology recently set up a Center for the Study of Ethics in the Professions.[56] These are new additions to academic curricula. If humanity is a product of molecular pushes and pulls, there is no reason why people should be concerned about the moral and ethical values of life. But every sensible person knows that there is value in life. Life per se is full of meaning and full of purpose.

Experimentally Observed Facts Concerning Life

The study of metabolic cellular pathways and the synthesis of a large number of molecules found in the living machinery have so far demonstrated that there is still something missing in our understanding of life. The synthesis of amino acids from a mixture of methane, ammonia, hydrogen, and water molecules by electric discharge, as reported by Miller, represents a merely chemical process. It in no way comes close to solving the riddle of the origin of life. Yet excitement over this simple experiment is so great that those who profess belief in chemical evolution have concluded that such steps are the ones that will finally lead to the evolution of a living cell. We do not, however, even know what a cell really is in its complete detail. It has been estimated that there may be as many as some 200 trillion molecules in a single cell, all executing thousands of coordinated reactions with precise timing and function.[57] Each step is performed in a specific order, keeping clear of other steps so as not to upset the balance of the reactions.

An example is the biosynthesis of L-isoleucine from L-threonine. We have noted that if the end product is supplied from an outside source, the synthetic steps are immediately stopped, the first step having been inhibited by the binding of the enzyme L-threonine deaminase with L-isolecine. This is referred to in biochemical language as a feedback inhibition mechanism. Likewise, in the transcription of DNA to RNA, and the translation of RNA to proteins, all the steps follow directed instructions. What makes a living cell perform all these seemingly purposeful chemical reactions? What are the chemical theories or principles that can explain such apparently conscious acts even at molecular levels? What is the wave function that can explain such phenomena?

One of the authors asked Stanley Miller, the molecular evolutionist, during one of his series of lectures at the University of California, Irvine, on the origins of life, "Suppose you were given all the necessary cellular chemicals. Could you create a living cell?" His immediate answer was, "I do not know."[58] The point is that if this experiment cannot be demonstrated, molecular evolutionists cannot honestly claim that life has arisen from molecules. As explained earlier, a molecule, no matter how orderly and precisely arranged, is lifeless. To make the artificial gene work, the help of a living cell is required. All the molecules, including DNA, are only vehicles to carry out an instructed message, just like the running of a watch. But are the watch and the watchmaker the same?

Let us now turn to some explicit contemplative questions. What is that molecular operation that makes us appreciate a beautiful landscape or listen to a nice symphony orchestra? What is that molecular operation which makes us feel joy upon seeing a close friend or relative after a long time, or sad when losing a near and dear one? What is that molecular operation which makes a squirrel sense its ability to jump from one branch of a tall pine tree to another with perfect timing and accuracy? What is that molecular operation which makes the Pacific northwest salmon undertake the mysterious and dramatic odyssey of swimming upstream hundreds of miles in the face of many obstacles just to spawn and then die? What is the molecular operation which directs the tiny sandpiper to follow the course of a subsiding wave along the seashore with hundreds of quick steps to find its food? What is that molecular operation which makes the cuckoo lay its eggs in the nests of other birds as a meaningful trick? What enables the Nile crocodile, whose jaws can crush the femur of a buffalo, to pick up its egg gently enough not to break it, freeing the hatchling without harming it?[59] What are those molecules which discriminate between such contrasting but spontaneous conscious acts? And finally, what is that molecular operation which makes thoughtful scientists come together to discuss the value of knowledge and the goal and meaning of life? Are all of these due to molecules? Is there any molecular operation or any multidimensional quantum mechanical equation that can describe these wonderful phenomena of life?

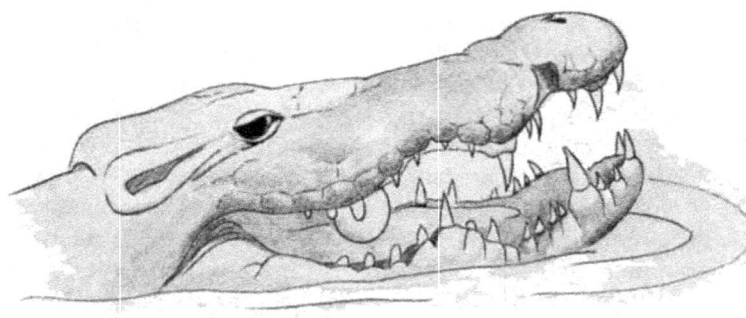

Figure 12. A Nile crocodile freeing its young from the egg.

36

On the human level there are so many subtle traits of personality—for example: humility, stability and self-control, honesty, tolerance, responsibility, cleanliness, love and so on. Are there any molecular mechanisms that can turn off and on to produce all these unique symptoms?

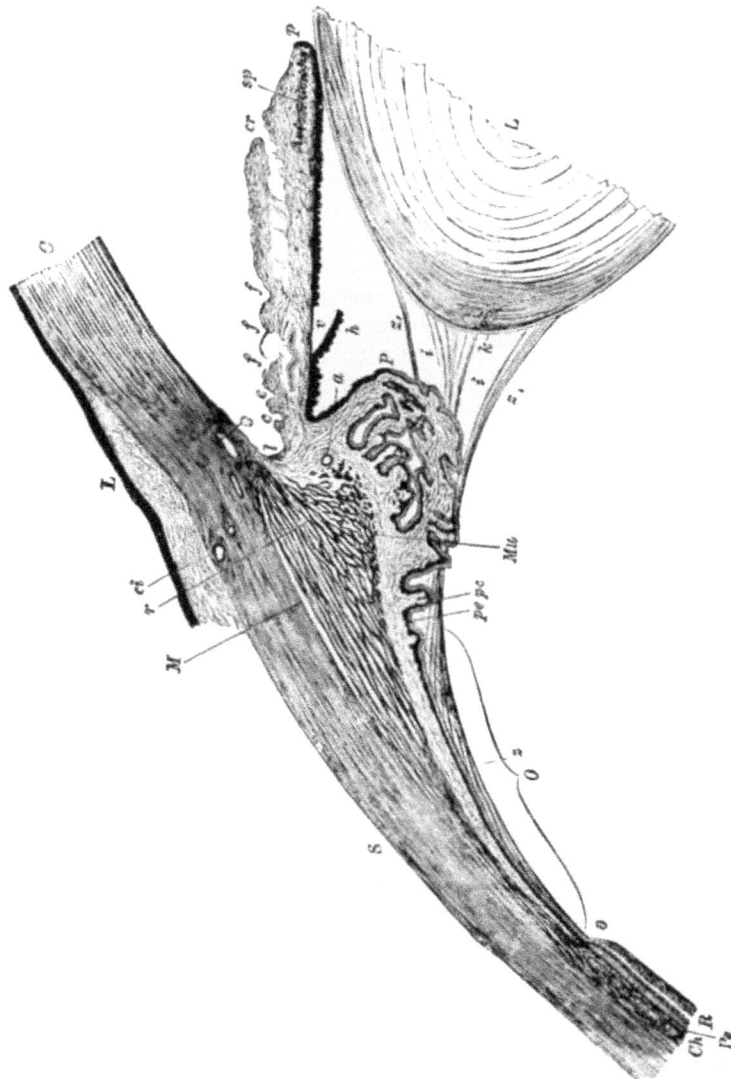

Figure 13. The structure of a human eye.

There are innumerable examples one may cite. We encounter marvels of life on so many levels, and the theorists of evolution cannot even think of touching these points. Darwin himself encountered insurmountable difficulty in conceiving how an eye could evolve. The fine, intricate details of the colorful feather in a peacock's tail were also impossible for him to explain. He thus remarked:

> I remember well the time when the thought of the eye made me cold all over, but I have got over the complaint, and now small trifling particulars of structure often make me very uncomfortable. The sight of a feather in a peacock's tail, whenever I gaze at it, makes me sick.[60]

Sometimes in human experience mental events happen suddenly and without apparent antecedents. Fine poetry comes from a poet's thought; the solution to a difficult mathematical riddle is revealed like a flash in the mind of a mathematician; an intricate chemical structure is revealed in the mind of a chemist; a whole symphony is inspired in the mind of a composer. Consider the experience of the famous composer Mozart:

> When I feel well and in good humor thoughts crowd into my mind as easily as you could wish. Whence and how do they come? I do not know and I have nothing to do with it. . . . Once I have my theme, another melody comes, linking itself with the first one, in accordance with the needs of the composition as a whole . . . Then my soul is on fire with inspiration, if however nothing occurs to distract my attention. The work grows: I keep expanding it, conceiving it more and more clearly until I have the entire composition finished in my head though it may be long . . . It does not come to me successively, with its various parts worked out in detail, as they will be later on, but it is in its entirety that my imagination lets me hear it.[61]

Are we to suppose that these phenomena are nothing but the products of chance and simple pushes and pulls? In Mozart we see a unique ability not present in any other members of his family. (Mozart's father, for example, was an ordinary musician.) If a biochemical machine was present in Mozart's brain that could generate symphonies effortlessly, where did this machine come from? If a human being were to design such a machine, he would certainly have to adjust many delicately interrelated variables, and this would require great intelligence and perserverance. Are we to suppose, then, that a random mutation of a gene or a chance combination of genetic alleles was able to produce such a machine? (We should note that the chance that a pattern will form randomly goes down exponentially with the number of variables entering into the pattern.) Or are we to suppose that by Coulomb's law and the spin-orbit interactions, such a machine will just naturally pull itself together, given enough time?

Limitations of the Laws of Nature

There is every reason to believe that simple push-pull laws are insufficient to account for all the effects of nature. The idea that they are known to be sufficient can easily be seen to be an illusion, for the quantum mechanical equations of Figure 2 can be solved exactly only in very simple situations. Even the calculations of the equilibrium state of a diatomic hydrogen molecule are very difficult and can be approached only by approximations requiring extensive computer calculations.

Yet the diatomic hydrogen molecule has only four particles—two electrons and two protons. For larger numbers of particles, practically nothing can be said about the solutions to the quantum mechanical equations. Glycine, the smallest amino acid, has 85 particles (if we count each nucleus as a unit), and cellular proteins contain hundreds of amino acids. It is therefore nothing but an article of faith that only the known laws of physics can account for life.

In addition, if we scrutinize the mathematical system of modern physics, we find serious contradictions and discrepancies suggesting that these laws are not adequate to describe even the phenomena of inanimate matter. These contradictions involve the most basic features of the theory of quantum mechanics, and concern the relation between its deterministic equations (the Schrödinger equation) and its statistical interpretation.

To enter into these matters here would take us too far afield. (An extensive treatment of this subject is provided in Monograph 3 of this series.[62]) However, we should note that many eminent scientists have held strong reservations about this theory. Einstein in particular felt that the theory was quite unrealistic. He remarked:

> The Heisenberg-Bohr tranquilizing philosophy—or religion?— is so delicately contrived that, for the time being, it provides a gentle pillow for the true believer from which he cannot very easily be aroused. So let him lie there.[63]

It is thus premature for biologists and biochemists to select this theory as the ultimate foundation for their understanding of life. In this section, we shall briefly present some mathematical findings directly indicating that simple laws cannot give rise to the complex structures we see in living organisms. These conclusions imply that the standard view of the laws of nature is incomplete; there must be many laws operating in nature which cannot be reduced to combinations of known physical interactions.

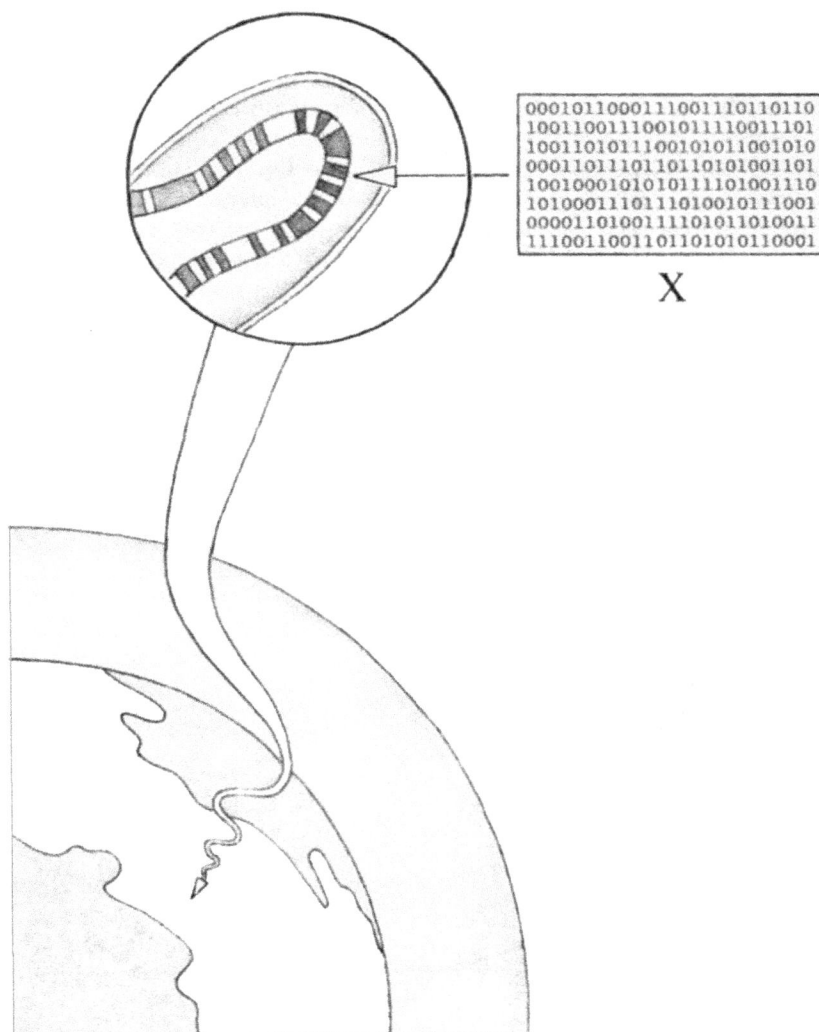

```
0001011000111001110110110
1001100111001011110011101
1001101011100101011001010
0001101110110110101001101
1001000101010111101001110
1010001110111010010111001
0000110100111101011010011
1110011001101101010110001
```

X

Figure 14. A biological structure coded by a binary sequence, X.

41

We shall give only a broad conceptual outline of these findings in this paper. (A detailed description is found in Monograph 2 of this series.) The essential idea is that simple laws lack the discriminating power to select highly complex forms from a welter of randomly distributed molecules. This can be expressed as follows in terms of the mathematical concept of information content:[64]

$$M(X) \leqslant 64^{60+L(M)-L(X)} \tag{5}$$

Here, $M(X)$ is the probability that the structure, X, will evolve somewhere on the earth in a 4.5-billion-year period (the estimated age of the earth according to current scientific opinion). $L(M)$ represents the information contained in the physical system itself. This includes the information contained in the fundamental natural laws, plus any information built into the initial state of the system, or entering it from outside. $L(X)$ represents the information content of the structure, X, which is being considered as a candidate for evolution.

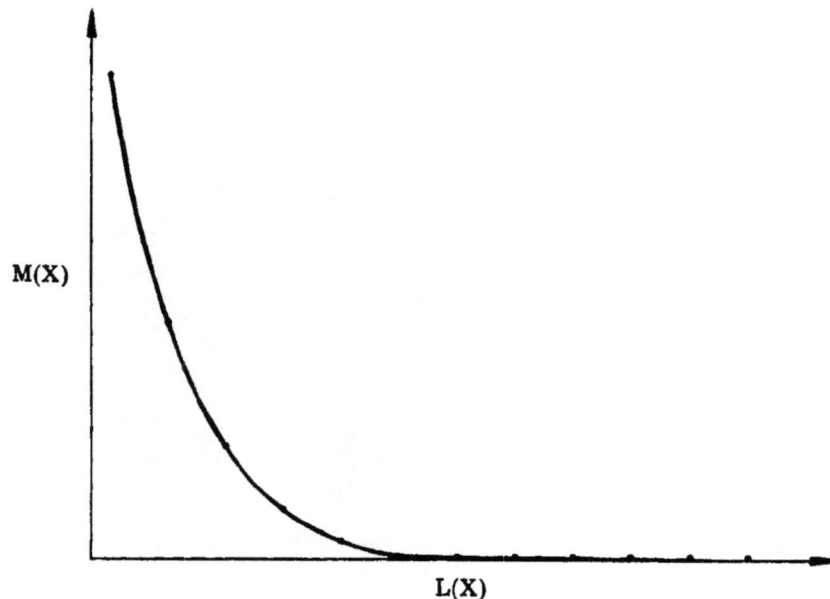

Figure 15. Limitation on the probability that X will evolve, as a function of its information content.

Earlier we considered the probability of forming a structure purely by chance, and saw that this probability went down exponentially (a^{-N}) with the number of variable elements (N) in the structure. Here the situation is similar, even though we are considering the simultaneous action of both random events and natural laws. The analysis in terms of information content indicates that all the information in X not supplied by the system must be provided by pure chance. Consequently, as the difference between L(X) and L(M) goes up, the probability of the evolution of X goes down exponentially.

Monograph 2 explains the concept of information content and argues that the elaborate structures of living organisms must embody large amounts of specific information. It follows that this information must come from somewhere. If the natural laws are very simple (as is true of the present laws of physics), then the information must come either from outside the system or from "chance." If it comes from chance, then the probability of evolution is circumscribed by bounds with large negative exponents (such as $64^{-80,000}$, as computed in Monograph 2). This means that evolution of complex life forms could not be expected to occur due to the action of chance and simple laws even in many billions of repetitions of the entire history of the earth.

The *E. coli* bacterium provides a practical example that may give some idea of why simple laws cannot be expected to generate the forms of living organisms. An *E. coli* cell possesses a number of spiral flagella. Each flagellum is connected by a kind of universal joint to a motor built into the outer wall of the cell.[65] When all the flagella are rotated in one direction by these motors, they act like the propellers of a submarine and drive the bacterium steadily through the water. The motors possess drive shafts and some kind of rotating discs, but the principle underlying their operation is still unknown.

Our question is: How did these motors originate? If we start with a cell that completely lacks such structures, then by what process could they arise? Surely the probability that one random mutation could produce the design for a complete functioning motor is exceedingly low, since many precisely structured components would have to be brought into existence at one time. On the other hand, if many successive mutations are required, then the standard theory of natural selection requires that each intermediate step be useful for the organism. But what useful

43

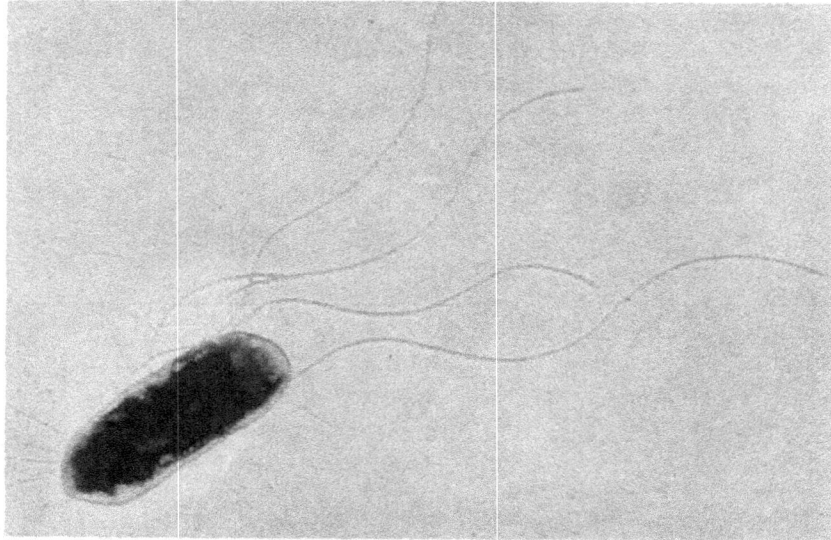

Figure 16. *Escherichia coli* bacterium.

intermediate steps are there between a motor and no motor? Can there exist a useful structure which is half a motor?

The process of natural selection is the only means envisioned thus far by which simple laws might yield complex organization. Yet if there does not exist a natural sequence of useful intermediate forms leading to a structure, this process cannot operate. Only pure chance is available to bridge the gaps, and this renders evolution fantastically unlikely, even over vast spans of time. Actually, there are innumerable instances of structures for which useful intermediate forms are hard to imagine. Also, the absence of the remains of intermediate forms in the fossil record has been one of the most striking findings of paleontology. This should give us some insight into why we should expect a limitation on evolution such as that expressed by inequality (5). (The mathematical derivation of (5) proceeds along somewhat different lines, however.)

This inequality implies that the information required to specify the structures of living organisms either must be built into the initial and boundary conditions of the system or must be built into the system's fundamental laws. To say that the necessary information was present in the initial state of the system certainly contradicts the idea of the disorganized primordial broth, or the earlier pre-solar gaseous nebula. It begs the question of origins by pushing it back to an earlier time, thus requiring some explanation for the origin of the complex initial conditions. The same can be said for the idea that the information was transmitted across the boundary of the system from outside in some form of code.

We, therefore, suggest that a more fruitful approach to the question of origins should begin with the understanding that the fundamental laws of nature are not as simple as modern scientists have supposed. It is to be expected that there are many additional laws specifically involved with life. It is interesting that in pondering the dilemmas of the quantum theory, the physicist E. Wigner was led to similar considerations. He proposed that

> The present laws of physics are at least incomplete without a translation into terms of mental phenomena. More likely they are inaccurate, the inaccuracy increasing with the increase of the role which life plays in the phenomena considered.[66]

A New Scientific Paradigm:
Laws Beyond the Laws of Nature

We have seen that life possesses qualities beyond the limits of our physical descriptions, in spite of all the claims of its origin from inanimate molecules. Another fundamental quality of life is consciousness. To our knowledge, molecular evolutionists have never seriously tried to explain consciousness, because the symptoms of conscious awareness are simply beyond the realm of molecular description. Here we encounter a strong drawback in the chemical model of life.

Out of frustration, some people intentionally try to neglect this. For example, Niels Bohr remarked: "An analysis of the very concept of explanation would naturally begin and end with a renunciation as to explaining our own conscious activity."[67] This again is intellectual dishonesty. Bohr tried to explain everything by the quantum theory. However, since he felt that consciousness could not be explained by this theory, he had no choice but to "renounce" it.

But consciousness exists nonetheless. As Wigner remarked, "thought processes as well as consciousness are the primary concepts, . . . our knowledge of the external world is the content of our consciousness, and . . . this consciousness therefore cannot be denied."[68] If we are to understand the mystery of consciousness, and the many other mysteries of life, it is clear that we cannot remain within the narrow confines of mechanical and molecular thinking. A broader perspective on reality is needed.

In this section we would therefore like to introduce an alternative view of the basic principles underlying nature. We have referred to these basic principles as the absolute truth, or the ultimate cause of all phenomena. Even though most scientific theories deal in practice only with relative descriptions of nature, the goal of science has always been to seek out the ultimate principles underlying reality. Yet certain far-reaching assumptions about these principles have provided the foundation for all modern scientific research. The dominant scientific view of the past two hundred years has been that these ultimate principles consist of a few basic natural laws which can be expressed by mathematical formulas. We have listed one current version of these laws in Figure 2.

As this view appears to be far too restrictive to account for the phenomena of life, we are going to propose a diametrically different view, which may provide a framework and an inspiration for further scientific research. This is essentially the view of the absolute truth as presented in the ancient Sanskrit text *Bhagavad-gītā.*[69] We would like to stress that this view is not being offered as a dogma or as a metaphysical explanatory device incapable of scientific test. Although many of its features may appear difficult to verify empirically, others have very direct implications concerning what we may expect to observe. This view should serve as a stimulating challenge to the truly sicentific spirit that wishes to go beyond the very restrictive framework imposed on our scientific understanding of nature for the past two hundred years.

In both of these viewpoints the absolute truth may be described as the ultimate cause, or causes, lying behind all the phenomena of nature. Figure 17 contrasts the two views.

The first three points indicate features that both views of the absolute truth hold in common. The first point is that the ultimate laws must exist in a fashion not fully comprehensible to the

Basic Features	The View of Modern Science: The Laws of Nature	The Alternative View: *Paramātmā*
1. The absolute truth exists, but it is not fully conceivable to the human mind.	Yes	Yes
2. It exists invariantly throughout space.	Yes	Yes
3. It does not change with time.	Yes	Yes
4. It controls all manifestations.	Yes*	Yes
5. It exists as a unified whole.	No	Yes
6. It possesses the attribute of consciousness.	No	Yes
7. It corresponds with fixed mathematical expressions.	Yes	No
8. Amount of inherent variety, or information content.	Small	Unlimited

*The natural laws of science must be supplemented by initial and boundary conditions in order to completely determine the course of events.

Figure 17. Two alternative views of the absolute truth.

47

human mind. This is illustrated, for example, by the law of gravity. We cannot imagine how a force can act across empty space to pull one object towards another, and yet the law of gravity postulates that such a "force" exists. For this reason, the law of gravity, when first proposed by Newton, was rejected as "occultism" by Leibnitz and other European philosophers. We can see, however, that a law must have some unexplainable features if it is actually fundamental: if the law can be explained in terms of other laws, it cannot, by definition, be fundamental.

Points two and three are also characteristic of both views, and these also represent inconceivable features. In science, a natural law is taken by definition to be invariant with respect to both space and time. If it were not invariant, then one could inquire by what law it varies, and that law would be taken instead as the fundamental law.

Point four should ideally be "yes" in both columns. We should expect the ultimate cause to determine all phenomena completely. The natural laws of modern science, however, must be supplemented by initial conditions describing the state of affairs in nature at some arbitrary point in time. This is a rather unsatisfactory feature of the modern scientific view, and theories such as the Darwinian theory of evolution and the "big bang" theory of cosmology may be viewed as attempts to circumvent it.

For example, if we were forced to account for the existence of life forms by postulating initial conditions in which life forms already existed, then we could hardly say that our natural laws had explained life. The theory of evolution avoids this by positing a natural mechanism whereby life forms could arise from a chaotic cloud of gas or a "primordial soup." In this way the required initial condition is rendered as simple as possible, and all significant phenomena are attributed to the operation of the laws themselves. As we have already pointed out, however, this theory cannot be expected to hold true: it is absurd to suppose that simple pushes and pulls alone could organize a chaotic, seething mass of atomic particles into a system of life forms capable of exhibiting so many remarkable qualities and activities.

Another feature of the modern scientific view is chance, which enters the theory of quantum mechanics as a kind of repeated initial condition in the so-called "reduction of the wave packet." The role of chance in modern physics has many highly unsatisfactory features that we shall not enter into here. The basic point is that chance enters modern physical theory as an arbitrary yet unavoidable correction factor that modifies the behavior of the system under the natural laws. It is thus another aspect in which the natural laws fail to completely specify the phenomena of nature.

These drawbacks of the modern scientific view suggest the existence of natural laws of a higher order. Such laws would serve to provide the missing information needed to account for the origin of life, and would also serve to fill in the missing causal determination represented by "chance" in modern physics. By "higher order" we shall refer to one of the following set of progressively stronger properties:

(1) The laws cannot be reduced to the known laws of physics and chemistry.
(2) They can be expressed mathematically only by very elaborate formulas.
(3) They cannot be expressed mathematically at all, and relate to entities not amenable to numerical description.

The problems of evolutionary theory and quantum mechanics at least call for the existence of high-order laws of types (1) and (2). The simple push-pull laws of modern physics and chemistry are certainly inadequate to account for the phenomena of life, and the dilemmas of quantum theory suggest that they are not even adequate to account for the phenomena studied in physics. The very elaborate structures and activities manifested by living beings are particularly indicative of laws of type (2).

The phenomenon of consciousness, however, indicates that even laws of type (2) will not suffice to give a complete description of reality. Consciousness exists, and there is every reason to believe that it is qualitatively irreducible to mathematical description of any kind. An array of numbers, no matter how elaborate, can tell us nothing about a person's conscious awareness, even though it might describe the person's external bodily movements and electrochemical reactions with great accuracy. Therefore, if we are to entertain the idea of a complete description of reality, we must consider laws of type (3).

Our alternative view is based on the idea that the fundamental laws of nature must account for all phenomena, and that a cause must be at least as great as its effect in terms of information content. Thus we propose that an unlimited reservoir of fundamental laws lies behind nature, and that they determine all the features of nature, including living organisms. The existence of such higher laws and principles clearly provides unlimited possibilities for future scientific investigation, investigation which should prove to be much more fruitful than the many past invocations of the marvelous powers of "natural selection" and "chance."

Points 5, 6, and 7 also go together since for these points the two views are opposed. One of the basic ideals of modern science has been to find unity in nature. Since the absolute truth has been viewed as a system of mathematical laws, this ideal has been expressed by the requirement that the basic equations have the greatest possible mathematical simplicity. Einstein was one of the strongest proponents of this goal, and it is epitomized by his search for a "unified field theory" which would derive all the laws of physics from one basic mathematical rule.

However, this unity has not been attained. The basic laws are simply a list of apparently unrelated formulas and expressions, and further disunity is provided by the initial conditions (as well as by the arbitrary entrance of "chance" into quantum mechanics). The existence of many irreducible higher-order laws would seem to suggest that the absolute truth has an even greater disunity and is analogous to a gigantic cosmic laundry list.

In our alternative view, however, this unity is provided by a higher principle extending beyond the realm of laws that are capable of mathematical formulation. This is the principle of consciousness. As we have indicated, consciousness has eluded scientific explanation, and we propose that it cannot, even in principle, be reduced to a mathematical formulation.

We should stress here that the postulate that nature is mathematically describable in all essential features is also a drastic and highly restrictive a priori assumption. Why should we expect that reality can be encompassed by the patterns of finite symbol manipulation that we can invent and contemplate with our limited minds? It is perfectly possible for an entity to exist that cannot be described by equations, even though it may exhibit many features that can be so described.

In our alternative view, consciousness is taken as a fundamental feature of the absolute truth, and all the basic laws and principles of nature are seen to be integrated into a harmonious whole within the awareness of absolute consciousness. This means that the absolute truth exists as one unified, sentient being. Such a statement may appear to lie outside the realm of experimental observation. We introduce it both for the sake of philosophical completeness and for its implication that we should expect to find higher-order laws of nature that are of a psychological character. Such laws make sense in the context of the absolute truth as a primordial conscious being, but they do not fit sensibly into the mathematical framework of modern science. There the presence of such laws makes the world appear like a puppet show, with an elaborate script but neither an audience nor an author.

Point 8 is quite significant. The modern scientific view tries to depict nature in terms of a reduction to simple entities: atoms, molecules, and so on. This implies that the absolute truth is severely limited. This arbitrary a priori constraint on the nature of the absolute truth is one of the primary reasons why modern science cannot explain life.

What Is Matter and What Is Life?

We shall see some direct implications of this thesis once we have considered the fundamental nature of individual living beings. This is outlined in Figure 18.

Matter	Life
1. The inferior energy of the absolute truth.	1. The superior energy of the absolute truth.
2. Satisfies conservation of energy.	2. Satisfies conservation of energy.
3. Eternal.	3. Eternal.
4. Obeys the laws of physics and chemistry to some extent.	4. Non-physical and non-chemical.
5. Lacks consciousness and inherent meaning and purpose.	5. Possesses consciousness and inherent meaning and purpose.

Figure 18. The distinction between life and matter.

First of all, in the alternative view we are describing, matter and life are understood to be two distinct kinds of energy. Life is designated as the superior energy because it possesses the fundamental feature of consciousness, whereas matter does not. Both of these energies are eternal, and both are composed of basic elemental units. Both satisfy principles of conservation similar to those which are familiar from modern physics.

Matter is essentially an insentient substrate from which temporary forms can be constructed by atomic combinations. It derives its properties from the absolute truth, and its transformations are governed by laws emanating from this source. However, it is qualitatively inferior to its source of emanation, since it lacks the inherent property of consciousness.

Life consists of innumerable fundamental units, which may be referred to as *ātmās*, or living entities. These are described in Figure 19. The *ātmā* may be thought of as a fundamental quantized part of the absolute living being possessing the irreducible property of consciousness. The *ātmā* may thus be compared to the electron, which is regarded as the fundamental quantum of electricity. These quanta of life share the qualities of their absolute source—including consciousness and purposefulness—in minute degree, and are thus regarded as the superior energy of the absolute truth.

The *ātmā:*

1. Is the quantum of life.
2. Exists in unlimited numbers.
3. Cannot be created or destroyed (conservation principle).
4. Possesses the property of consciousness and free will.

Figure 19. Properties of the *ātmā.*

Both life and matter operate according to the same natural laws, or ultimate causative principles. However, certain laws are more specifically associated with life, and others are more specifically associated with matter. The simple push-pull laws of physics and chemistry undoubtably have some bearing on the behavior of matter, especially in circumstances where life is not significantly involved (inanimate matter). However, these are at best limiting cases of more general laws which are involved with life.

The interaction of life with matter ultimately depends upon higher-order principles that cannot be reduced to mathematical formulation. Essentially, the conscious, superior energy interacts with the inferior energy through the consciousness of the absolute truth. This interaction cannot be completely described in quantitative terms, but it can be understood and investigated. It entails fundamental psychological principles such as free will, purpose, and value. Ultimately, this interaction can be understood as the direction and supervision of the individual *ātmās* by the *paramātmā:* as the individual *ātmās* develop various desires and psychological states in the course of their experiences, the *paramātmā* observes these and adjusts the material situation accordingly.

Thus, the distinction between matter and life is the quality of consciousness. This is the main reason why scientists have had such difficulty in defining life. They either try to avoid consciousness completely, or they try to imagine generating it by molecular combination of inanimate matter.

Matter Under the Influence of Life

The actual subject of study in modern biology is thus not life itself, but the material structures that are associated with life and that develop according to the laws governing the interaction of life and matter. In Figure 20 we have outlined some specific indications of the influence of life on matter in familiar living organisms. The distinction between the two categories is, of course, a relative one, for the qualities of the materially conditioned living entities can be expressed through material arrangements to varying degrees.

First of all, matter by itself does not tend to exhibit very much specific information content. It is either found in simple organized forms like the diamond crystal, or it lacks organization altogether. On the other hand, the structures of living organisms exhibit an intricate organization which we are just beginning to understand. Consider the many complex systems involved in the human eye alone, for example. As we have already pointed out, this complexity strongly suggests that higher-order laws are involved.

Secondly, matter by itself tends to reduce to thermodynamically stable forms that usually consist of small molecules exhibit-

Matter by itself	Matter in the presence of life
1. Characterized by either low information content or absence of specific form.	1. Characterized by high information content and very specific form.
2. Reduces to thermodynamically stable states.	2. Thermodynamically unstable states play a dominant role.
3. No highly organized flow of matter.	3. Exhibits a precisely regulated flow of matter.
4. Tends to lose form or pattern under transformation.	4. Undergoes transformation without loss of complex pattern (reproduction).
5. Grows by external accumulation only.	5. Grows from within by an intricate construction process.
6. Exhibits only passive resistance.	6. Adaptive: tries to actively overcome obstacles.

Figure 20. The influence of life on matter.

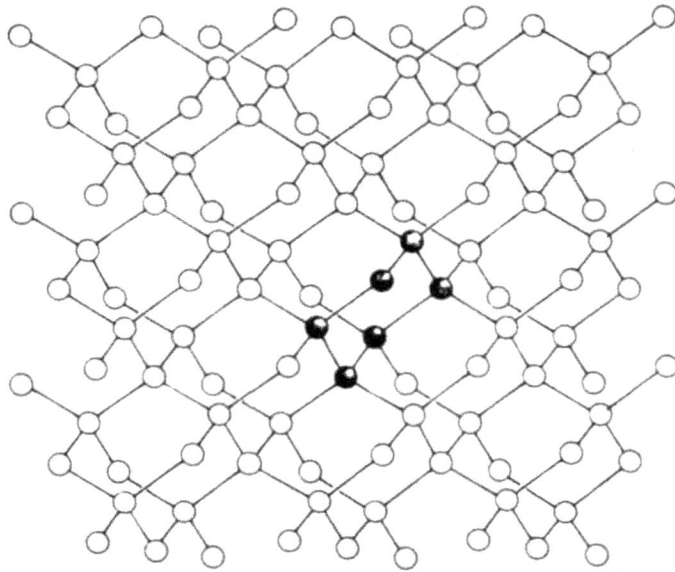

Figure 21. Diamond crystal lattice.

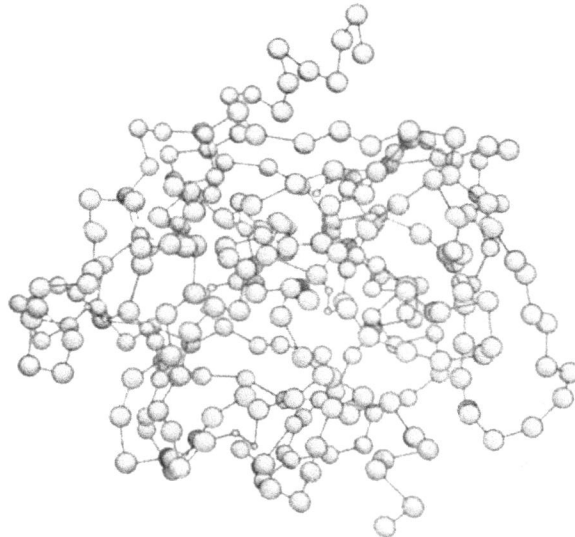

Figure 22. The chymotrypsin molecule.

55

ing little activity. On the other hand, in living organisms we see highly unstable molecules, such as the molecule of chymotrypsin illustrated in Figure 22. Such molecules are very readily broken down or denatured when subjected to ordinary chemical reactions.

Matter by itself tends to exhibit very simple patterns of flow, as we see for example in the flow of a river to the sea. Within living organisms, however, we see the kind of highly regulated chemical processes shown in Figure 23. This figure illustrates a

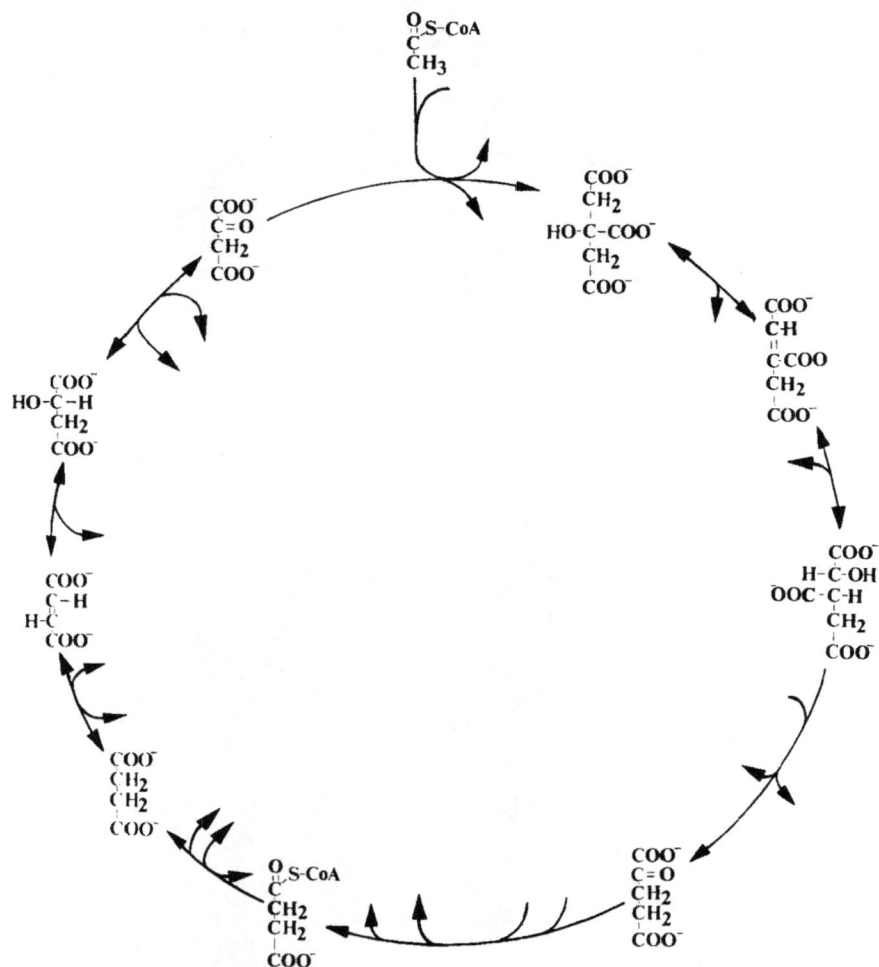

Figure 23. The flow **of matter in human metabolism.**

sequence of cellular chemical reactions known as the Kreb's cycle. Each reaction is controlled by a specific enzyme, and is adjusted so that the precise amount of the product needed by the cell is produced. The entire metabolism of even the simplest bacterial cells must contain thousands of precisely coordinated reactions of this kind. Thus, even if the cell is simply a chemical automation, it must have built into it logical instructions of complexity and sophistication that far surpass any computer program yet written by a human being. We propose that the higher order interactions ultimately stemming from the *paramātmā* play a role in the functioning of a cell analogous to the role of the programmer in a man-made computer. (In this connection we should note that computer programs are notorious for the amount of "debugging" work required to get them to work properly and to continue working in novel circumstances.)

Reproduction is another feature of living organisms that distinguishes them from inanimate matter. When inanimate matter is transformed, it tends to lose whatever structure or organization it may once have had. Consider, for example, the gradual decay of an abandoned car. In contrast, living organisms everywhere exhibit the renewal of their kind without apparent loss in vitality. We may also note that there are differences in the patterns of growth exhibited by inanimate matter and matter under the influence of life. In the former, growth proceeds by simple accumulation, as

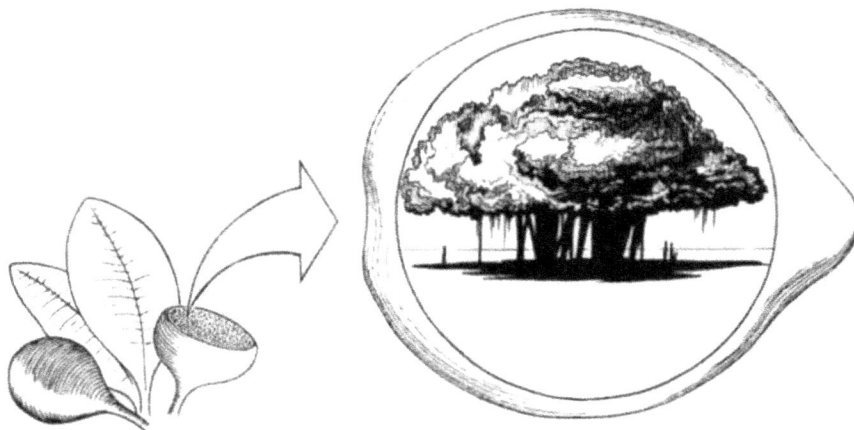

Figure 24. The process of perpetual renewal of living organisms.

57

we can see in the growth of a crystal. In a living organism, however, growth occurs by an elaborate internal construction process. At present, scientists have only dimly surmised the principles underlying such growth processes as the development of an embryo from a fertilized egg.

Finally, we can see that living organisms adapt and actively strive in many ways to overcome obstacles or achieve goals. This is exhibited in the process of healing, the migration of birds, the climbing of mountains, the building of industrial empires, and innumerable other examples. However, inanimate matter exhibits only passive resistance to change. Even man-made computers fail to display the kind of flexible adaptability visible in so-called primitive organisms. (In fact, computer systems tend to go out of control very easily, despite all attempts to build in safeguards, and for this reason they require constant human attention.)

Scientific Methods of Studying Matter and Life

We suspect that there are many fruitful avenues of scientific research that will lead to an understanding of some of the specific laws involved with life. We would particularly like to address the direct empirical investigation of the living entity, or *ātmā*, itself. The most significant implication of the new paradigm we have outlined is that it admits the possibility of the *ātmā's* recognizing its own existence, apart from matter. Figure 25 suggests how this investigation should be approached.

The basic idea here is that a material entity, such as an electron, is normally studied by taking advantage of its characteristic properties and laws of interaction. For example ESR spectroscopy takes advantage of the electron's magnetic interaction (or "spin" interaction) by the use of an oscillating magnetic field. If an entity does not interact in a particular way, then it is, of course, useless to try to observe it using that type of interaction. (For this reason the particle known as the neutrino is very hard to study—it does not interact very strongly with any of the standard measuring instruments employed by physicists.)

Similar considerations apply to the *ātmā*. According to the alternative view that we are presenting, the *ātmā* interacts with matter through the agency of the *paramātmā*, or the all-pervading conscious aspect of the absolute truth. This interaction depends

	MATTER	LIFE
Entity	Electron	*Ātmā*
Property	Charge	Consciousness
Laws of Interaction	Coulomb force law, Spin interaction	Psychological laws mediated through absolute consciousness
Means of Experimental Measurement	Set up an apparatus to take advantage of this law of interaction.	The same principle: we must also take advantage of the laws of interaction.
Example	Electron spin resonance spectroscopy	The techniques of *bhakti-yoga*

Figure 25. The study of life and matter contrasted.

59

on higher-order, non-mathematical laws relating to psychological principles such as desire and free will. For this reason we should not expect to be able to observe the *ātmā* by means of a standard physical apparatus, such as an electron microscope, which employs the familiar physical laws. However, since both the *ātmā* and the absolute truth, or *paramātmā*, are conscious, there is the possibility of direct conscious interaction between them. It is through this interaction that the *ātmā* can be directly studied, and this study also entails the study of the *paramātmā*.

Even though this subject matter is unknown in the domain of Western scientific knowledge, systematic and scientific procedures for the study of the *ātmā* have existed for a very long time. Centuries ago, they were expounded in the *Bhagavad-gītā* and other Sanskrit texts, and more recently they have been treated in great detail in such works as the *Bhakti-rasāmṛta-sindhu* of Śrīla Rūpa Goswāmī.[70] Here we will only give a brief outline of some basic principles underlying these procedures. In subsequent monographs in this series they will be described in detail.

Essentially, the object of study in this investigation is the investigator's personal self. The stringent experimental conditions necessary in ordinary physical experimentation must, therefore, be applied to the mind and senses, rather than to an external experimental apparatus. The sensory apparatus for the study of the *ātmā* resides in the *ātmā* itself. In the materially conditioned state the *ātmā* normally perceives external arrangements of matter through the gross senses of the physical body. The data for such perception pass through sense organs, nerves, and other physical structures. But in order for perception to take place, these data must ultimately reach the *ātmā*, or quantum of consciousness. The *ātmā* must therefore possess its own senses, or means of direct perception. It is this capacity of direct perception which must be invoked in order for the *ātmā* to study itself, other *ātmās*, and the *paramātmā*. Perception through the limiting medium of gross material instruments, including the bodily senses, can only give indirect evidence of the *ātmā*, since these instruments make use of an inferior mode of interaction.

One of the first steps required for the study of conscious interaction is therefore to purify the mind of the materially conditioned *ātmā*. A materially conditioned *ātmā*, or soul, is one who, devoid of real knowledge, assumes that he is a product of material interactions. He thinks that everything is within his power, and has a mental attitude characterized by the desire to lord it over nature. He tends to think that he can shape his own destiny, and that there is nothing beyond chance and matter. The words of the Nobel-prize-winning scientist Jacques Monod are indicative of this type of mentality:

> The ancient covenant is in pieces; man knows at last that he is alone in the universe's unfeeling immensity, out of which he emerged by chance. His destiny is nowhere spelled out, nor is his duty.[71]

Unfortunately, this attitude makes progress in the direct study of life extremely difficult.

As we have pointed out, the study of life ultimately involves the interaction between the individual quantum of consciousness and absolute, or universal consciousness, *paramātmā*. For the minute conscious entity to approach the supreme conscious source of all entities, a reverence for life in all its forms is needed, as well as a desire to cooperate harmoniously with the absolute source from which all life emanates. The study of life thus requires that higher psychological laws must be taken into account, just as the study of gross matter requires an appreciation of the natural laws which predominate in each particular experimental situation.

Conclusion

In this monograph we have pointed out that the theory of molecular evolution is not scientifically valid. This inherently weak theory has arisen because its propounders have no clear understanding of life and matter. To them life is just a mechanical transformation of inanimate matter, and they cannot speak about life in any language except that of chemistry and physics. But we have indicated that this approach is incompatible with the observed facts. We have further shown how life and matter can be understood as two completely different categories. One is not reducible to the other, although the latter can be transformed into structures of different sizes, shapes, and colors by the influence of the former.

Since life is a non-physical and non-chemical entity, any attempt to understand life in terms of chemistry and physics cannot go very far. Honest and thoughtful scientists are beginning to realize this.

We have discussed an alternative scientific viewpoint. Both the modern scientific approach and the new paradigm agree that there is an absolute truth. However, the view of modern science reduces the absolute truth to nothing but some pushes and pulls of interacting atomic particles. This view is very unsatisfying and cannot meaningfully explain many observed phenomena pertaining to both life and matter. The new paradigm, however, reveals that the absolute truth is a supremely conscious being, identified as *paramātmā*. *Paramātmā* possesses unique features or qualities for generating both matter and life. As the leader of a nation is different from his functionary departments, although they are dependent upon him, so both life and matter emanate from that supremely conscious being, the original life, although they are different energies. This is quite reasonable, and it can explain all the features of both life and matter, as well as open up new possibilities of investigation precluded by the reductionistic view. We suggest that serious consideration of this new scientific paradigm will prove very fruitful. The famous scientist Louis Pasteur remarked:

> I have been looking for spontaneous generation for twenty years without discovering it. No, I do not judge it impossible. But what allows you to make it the origin of life? You place matter before

life and you decide that matter has existed for all eternity. How do you know that the incessant progress of science will not compell scientists ... to consider that life has existed during eternity, and not matter? You pass from matter to life because your intelligence of today ... cannot conceive things otherwise. How do you know that in ten thousand years one will not consider it more likely that matter has emerged from life?[72]

Notes and References

1. Paecht-Horowitz, M. *Agnew. Chem. Internat. Edn.*, 12, 349 (1973).
2. Lemmon, R.M. *Chem. Rev.*, 70, 95–109 (1970).
3. Ferris, J.P. and Chen, C.T. *J. Amer. Chem. Soc.*, 97, 2962 (1975).
4. Horowitz, N.H. *Acc. Chem. Res.*, 9, 1 (1976).
5. Darwin, F. *The Life and Letters of Charles Darwin*, Vol. 3. London: Murray, 1887, p. 18.
6. Monod, J. *Chance and Necessity*, trans. A. Wainhouse. New York: Alfred A. Knopf, 1971, pp. 112-113.
7. Kervran, C.L. *Biological Transmutations*. Binghamton, New York: Swan House Publishing Co., 1972, p. 1.
8. Bell, E.T. *Men of Mathematics*. New York: Simon and Schuster, 1965, p. 172.
9. Woodward, R.B. and Hoffman, R. *The Conservation of Orbital Symmetry*. Verlag Chemie GmbH, Academic Press, 1970.
10. Schrödinger, E. *What is Life?* New York: Macmillan, 1945, p. 2.
11. Bohr, N. *Atomic Physics and Human Knowledge*. New York: John Wiley and Sons, 1958.
12. Heisenberg, W. *Physics and Beyond*. New York: Harper and Row, 1971, pp. 110-111.
13. Watson, J.D. *Molecular Biology of the Gene*, 2nd ed. Menlo Park, California: W.A. Benjamin, 1970.
14. Crick, F. *Of Molecules and Man*. Seattle and London: University of Washington Press, 1976.
15. Thompson, R.L. *Consciousness and the Laws of Nature*. Bhaktivedanta Institute Monograph Series, No. 3. Boston: Bhaktivedanta Institute, 1977.
16. This is the non-relativistic Hamiltonian.
17. Watson, J.D. and Crick, F. *Nature*, 171, 964 (1953).
18. Watson, J.D. *Molecular Biology of the Gene*, p. 261.
19. *Ibid.* p. 281.
20. *Ibid.* pp. 295-296.
21. *Ibid.* pp. 282-292.
22. *Chem. Eng. News*, 54 (No. 39), 27 (1976).
23. *Biology Today*. Del Mar, California: CRM Books, 1972, p. xxiv.
24. Oparin, A.I. *The Origin of Life*, trans. S. Morgulis. New York: Dover, 1938.
25. Urey, H.C. *Proc. Nat. Acad. Sci. U.S.*, 38, 1952.
26. Miller, S.L. *Science*, 117, 528 (1953).
27. Abelson, P.H. *Proc. Nat. Acad. Sci. U.S.*, 55, 1365 (1966).
28. Davidson, C.F. *Proc. Nat. Acad. Sci. U.S.*, 53, 1194 (1965).
29. Brinkmann, R.T. *J. Geophy. Res.*, 74, 5355 (1969).
30. Orgel, L.E. *The Origins of Life: Molecules and Natural Selection*. New York: John Wiley and Sons, 1973, p. 112.

31. Matthews, C.N. *Origins of Life*, 6, 155-162 (1975).
32. Oparin, A.I. *The Chemical Origin of Life*, trans. A. Synge. Springfield, Illinois: Charles C. Thomas, 1964.
33. Dose, K., Fox, S.W., Deborin, G.A., and Pavlovskaya, T.E., eds. *The Origin of Life and Evolutionary Biochemistry*. New York and London: Plenum Press, 1974, p. 119.
34. Oparin, A.I. *The Chemical Origin of Life*, p. 46.
35. *Ibid.* p. 108.
36. Fox, S.W. *et al., Chem. Eng. News*, 48, 80 (June 22, 1970).
37. Fox, S.W. *Chem. Eng. News*, 49 (No. 50), 48 (1971).
38. Orgel, L.E. and Miller, S.L. *The Origins of Life on the Earth.* Englewood Cliffs, New Jersey: Prentice-Hall, Inc., 1974, p. 144.
39. *Ibid,* p. 145.
40. Gore, R. *National Geographic*, 150 (No. 3), 390 (1976).
41. Watson, J.D. *Molecular Biology of the Gene*, p. 85.
42. Lehninger, A.L. *Biochemistry*, 2nd ed. New York: Worth Publishers, Inc., 1975, p. 234.
43. Watson, J.D. *Molecular Biology of the Gene.* p. 281.
44. In this formula we assume that the "primordial soup" is divided into boxes 20 x 20 x 25 $Å^3$ in size. Molecules can be created and destroyed in each box at a rate of one per microsecond. They can also jump from box to box at a rate of no more than one jump per microsecond. It is assumed that their jumping creation, and destruction are completely random. The term, $(10^{10}P)^k/k!$, is an estimate for the probability of finding at least k events out of 10^{10} opportunities, assuming that events occur randomly with probability P. There are less than $10^{10}/16$ "boxes" in the volume of a typical *E. coli* cell (10^{12} $Å^3$). The number, 1.6×10^{61}, is an upper bound for:

$$\frac{\text{(volume of ``soup'')} \times \text{(microseconds in } 10^{12} \text{ years)}}{\text{(volume of } E. \text{ } coli \text{ cell)}}$$

Formula (5) thus gives a bound on the probability that k specific molecules will be found together in some *E. coli* sized volume in the soup at some time in a 10^{12} year period. By taking faster rates for the molecular events one can obtain the same results for proportionately shorter time spans.

45. Sherman, I.W. & V.G. *Biology—A Human Approach.* New York: Oxford Univ. Press, 1975, p. 4.
46. Wald, G. *Adventures in Earth History*, ed. P. Cloud. San Francisco: W.H. Freeman and Co., 1970, p. 429.
47. *The Humanist*, Vol. XXXVII, No. 1, 1977, p. 4.
48. Crick, F. *Of Molecules and Man*, p. 93.
49. Huxley, A. "Confession of a Professed Atheist." *Report*, Vol. III, No. 9, June 1966, p. 19.

50. MacDonald, M.R. *The Origin of Johnny.* New York: Alfred A. Knopf, 1975.
51. *Biology Today.* Del Mar, California: CRM Books, 1972.
52. deBeer, G. ed. *Charles Darwin and Thomas H. Huxley, Autobiographies.* London: Oxford University Press, 1974, p. 83.
53. *Ibid.,* pp. 83–84.
54. (a) *Chem. Eng. News,* 51 (No. 2), 14 (1973).
    (b) Seltzer, R.J. *et al., Chem. Eng. News,* 50 (No. 3), 14 (1972).
    (c) *Chem. Eng. News,* 55 (No. 7), 6 (1977).
    (d) *Ibid.,* 55 (No. 11), 7 (1977).
    (e) Levitt, A.E. *Chem. Eng. News,* 49 (No. 42), 25 (1971).
55. *Chem. Eng. News,* 49 (No. 42), 7 (1971).
56. *Ibid.,* 55 (No. 8), 26 (1977).
57. Gore, R. *National Geographic,* 150 (No. 3), 358 (1976).
58. Professor S.L. Miller, the molecular evolutionist, gave lectures on "The Origin of Life" at the University of California, Irvine, on May 30–31, 1973. The questions were put by T.D. Singh.
59. Pooley, A.C. and Gans, C. "The Nile Crocodile." *Scientific American,* 234 (No. 4), 114–124 (April 1976).
60. Macbeth, N. *Darwin Retried: An Appeal to Reason.* Boston: Gambit, 1971, p. 101.
61. Hadamard, J. *The Psychology of Invention in the Mathematical Field.* Princeton: Princeton University Press, 1949, p. 16.
62. Thompson, R.L. *Consciousness and the Laws of Nature.* pp. 65–73.
63. Jammer, M. *The Philosophy of Quantum Mechanics.* New York: John Wiley and Sons, 1974, p. 130.
64. Thompson, R.L. *Demonstration by Information Theory that Life Cannot Arise from Matter.* Bhaktivedanta Institute Monograph Series, No. 2. Boston: Bhaktivedanta Institute, 1977.
65. Berg, H.C. "How Bacteria Swim." *Scientific American,* 233 (No. 2), 36–44 (August 1975).
66. Wigner, E.P. "Physics and the Explanation of Life." *Foundations of Physics,* 1 (No. 1), 44 (1970).
67. Bohr, N. *Atomic Physics and Human Knowledge,* p. 11.
68. Wigner, E.P. "Remarks on the Mind-Body Question." *The Scientist Speculates,* ed. I.J. Good. New York: Basic Books, 1962, p. 290.
69. His Divine Grace A.C. Bhaktivedanta Swami Prabhupāda. *Bhagavad-gītā As It Is.* New York: Macmillan, 1972. This is a complete scientific text describing the nature of life (*ātmā*). Our paradigm is to translate this description into modern scientific language.
70. His Divine Grace A.C. Bhaktivedanta Swami Prabhupāda. *The Nectar of Devotion.* Los Angeles: Bhaktivedanta Book Trust, 1970. This is a summary study of *Bhakti-rasāmṛta-sindhu* by Śrīla Rūpa Goswāmī.
71. Monod, J. *Chance and Necessity.* p. 180.
72. Koestler, A. *The Act of Creation.* New York: Macmillan, 1964, p. 702.

Svarūpa Dāmodara Dāsa Brahmacārī (Thoudam D. Singh) was born in a Vaiṣṇava family in Manipur, India in 1941. He did his undergraduate work at Gauhati University, and earned his master's degree in chemistry at Calcutta University. In 1967 he came to the United States of America and studied at Canisius College and the University of California, Irvine, where he finished his Ph.D. in Physical Organic Chemistry in 1974. From 1974 to 1977 he worked as a Research Fellow at Emory University, studying reaction kinetics.

In 1970 he met His Divine Grace A. C. Bhaktivedanta Swami Prabhupāda, and was greatly impressed by his spiritual stature and authority. He relates: "I vividly remember the beautiful loving exchange of flowers between His Divine Grace and the disciples early in the morning in front of the temple. I was quite amazed to see the beautiful Śrī Śrī Rādhā Kṛṣṇa Temple in Los Angeles for the first time. Later on, Rāmānanda Prabhu took me upstairs to see Śrīla Prabhupāda. I received the most affectionate and compassionate glance from Śrīla Prabhupāda, and immediately felt: Here is my spiritual father." He was initiated as a disciple of Śrīla Prabhupāda in 1971.

At present, he is GBC and director of Bhaktivedanta Institute. His research interests as an institute member include molecular biology and the origin of life, the nature of consciousness, and biomedical ethics.

Sadāputa Dāsa (Richard Thompson) was born in Binghamton, New York in 1947. He spent his undergraduate years at the State University of New York at Binghamton, and received his Ph.D. in mathematics from Cornell University in 1974. He has published a monograph on theoretical statistical mechanics as Memoir 150 of the American Mathematical Society.

Throughout his studies, he was struck by the lack of any meaningful foundation to reality in modern scientific theories. His dissatisfaction with this culminated in 1970, when he studied the reduction of man to a Turing machine, a kind of abstract clockwork. Surely, he felt, the truth must be something different from this. Consequently, he began to study many different philosophies, with a view to finding a practical route to higher knowledge. In 1972 he discovered some of the books of His Divine Grace A. C. Bhaktivedanta Swami Prabhupāda, and was struck by the beauty of their conceptions and the clarity of their presentation. Here was a deeply meaningful philosophy capable of practical application in day-to-day life. He became initiated as Śrīla Prabhupāda's disciple in 1975, and is now a member of Bhaktivedanta Institute.

Monograph 2

DEMONSTRATION BY INFORMATION THEORY THAT LIFE CANNOT ARISE FROM MATTER

The Bhaktivedanta Institute
Monograph Series Number 2

Demonstration By Information Theory That Life Cannot Arise From Matter

by
Richard L. Thompson
(Sadāputa Dāsa Adhikārī)

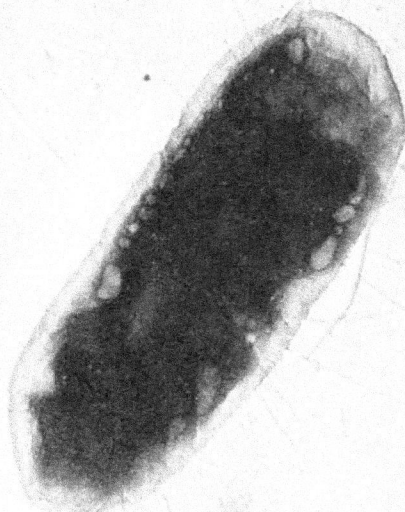

Demonstration By Information Theory
That Life Cannot Arise From Matter

Demonstration by Information Theory

That Life Cannot Arise From Matter*

by

Richard Thompson, Ph.D.

(Sadāputa Dāsa Adhikārī)

The Bhaktivedanta Institute Monograph Series   Number 2

*This monograph forms part of a forthcoming book, *The Origin of Life and Matter*, by Thoudam D. Singh, Michael Marchetti, and Richard Thompson.

iii

Readers interested in the subject matter of this monograph are invited to send correspondence to the Bhaktivedanta Institute at the following addresses.

70 Commonwealth Avenue
Boston, Massachusetts 02116
U.S.A. (617) 266-8369

Hare Krishna Land
Juhu, Bombay 400 054
India (Phone: 57-9373)

Bhaktivedanta Gurukula and
Institute for Higher Studies
Bhaktivedanta Swami Marg
Vrindavana, Mathura
India

v

Dedicated to His Divine Grace

A. C. Bhaktivedanta Swami Prabhupāda

om ajñāna-timirāndhasya jñānāñjana-śalākayā

cakṣur unmīlitaṁ yena tasmai śrī-gurave namaḥ

About Bhaktivedanta Institute

Bhaktivedanta Institute is a center for advanced study and research into the Vedic scientific knowledge concerning the nature of consciousness and the self. The Institute is the academic division of the International Society for Krishna Consciousness. It consists of a body of scientists and scholars who have recognized the unique value of the teachings of Krishna Consciousness brought to the West by His Divine Grace A. C. Bhaktivedanta Swami Prabhupāda. The main purpose of the Institute is to explore the implications of the Vedic knowledge as it bears on all features of human culture, and to present its findings in courses, lectures, monographs, books, and a quarterly journal, *Sa-vijñānam*.

The Institute presents modern science and other fields of knowledge in the light of Vaiṣṇava philosophy and tradition, providing a new perspective on reality quite different from that of our modern educational systems. One reason for the increasing interest of modern intellectuals in Śrīla Prabhupāda's teachings is doubtlessly the growing awareness that in spite of great scientific and technological advancements, the real goal of human life has somehow been missed. The philosophy of Bhaktivedanta Institute provides a meaningful answer to this concern by proposing that life—not matter—is the basis of the world we perceive.

The central doctrine of modern science is that all phenomena, including those of life and consciousness, can be fully explained and understood by recourse to matter alone. The dictum that "life is a manifestation of matter" is, indeed, the ultimate rationale for the entire civilization of material aggrandizement. The Vedas, on the other hand, teach that conscious life is original, fundamental, and eternal. This is the essence of *Bhagavad-gītā—"ahaṁ sarvasya prabhavo mattaḥ sarvaṁ pravartate."* (10.8) On this fundamental and critical point, modern science and Vedic knowledge find themselves opposed.

Bhaktivedanta Institute is dedicated to disseminating this most fundamental knowledge throughout the world. The Institute is clearly demonstrating that the Vedic version is not a matter simply of "faith" or "belief", but is scientific in the strict sense of the term. Although many of its features may appear difficult to verify experimentally, others have direct implications concerning what

we may expect to observe. Thus, this view should serve as a stimulating challenge to the truly scientific spirit to go beyond the very restrictive framework imposed on our scientific understanding of nature over the last two hundred years. Modern science began as an experiment to see how far nature could be explained without invoking God. But the purpose of Bhaktivedanta Institute is to introduce Vedic knowledge on a genuinely scientific basis for the first time in the history of this modern scientific age.

x

CONTENTS

Introduction

In this paper we will show how the mathematical analysis of information content demonstrates that the laws of nature as understood by modern science are not sufficient to account for the life we see about us. The basic argument is this: The laws of nature and the corresponding mathematical models of physical reality can all be described by a few simple equations and other numerical expressions. This means that they possess a low information content. On the other hand, the intricate and variegated forms of living organisms possess a high information content. It can be shown that configurations of high information content cannot arise with substantial probabilities, in models defined by mathematical expressions of low information content. It therefore follows that life forms could not arise by the action of the kind of natural laws or fundamental causal agents which are considered in modern science.

This argument has the particular feature of bypassing the proposition made by evolutionary theorists that even though the steps leading to life are improbable, they are nonetheless likely to happen, given the immense spans of geological time available. We show that no period of time from zero to billions of billions of years will suffice to make the evolution of life from matter by chance and natural selection a probable affair. Rather, we show that the probability of the evolution of life forms over periods many times that of the estimated 4.5-billion-year age of the earth is bounded by numerical limits of the order of $64^{-80,000}$, an almost infinitesimal number. This implies that the entire history of the earth would have to be repeated over and over again on the order of $64^{80,000}$ times in order for there to be a substantial chance that complex living entities would evolve even once.

An intuitive reason for the impossibility of organic evolution lies at the basis of these figures. It can be shown that the process of natural selection—the alleged mechanism of evolution—must have specific direction in order to bring about the evolution of complex living organisms. Without this direction, this process is unable to discriminate among random events (mutations) in such a way as to bring complex order out of chaos. For this reason, the standard argument that evolution will occur, given large enough time spans, is false. Natural selection lacks direction because the basic cause of this selection, namely the laws of nature, has too low a level of information content to specify such direction.

More generally, it is shown that in a mathematical system, any information which can be extracted from the system after transformations corresponding to the passage of time must have been built into the system in the first place. Random processes cannot contribute anything. It follows that mathematical systems are in principle unable to explain the origin of highly variegated and complex entities, such as the forms of living organisms. At best, such systems can merely account for complex order here and now by postulating that an equivalent complex order was present at some earlier time or was transported into the system from outside. This does not account for the origin of such order, but simply confronts us either with an infinite regress or with an eternal source of order containing the information necessary to specify all life forms.

This indicates that the original scientific program of describing the world by mathematical systems is too limited and can therefore only serve to impede our understanding of nature. This program is based on the conviction that the simple regularities observed by physicists and chemists in experiments with inanimate matter will suffice to account for every phenomenon in the world. But since we show here that this program cannot succeed in accounting for the origin of life, we suggest that the alternative view—that life, rather than matter, is the fundamental cause of all observed phenomena—must be seriously considered.

The program of science during the last two or three hundred years has been to reduce life to matter and deny the existence of any higher living principle transcendental to matter. In this program, the idea that matter is both simple and conceivable has been essential. But if, as shown here, it is necessary to attribute to matter all the characteristic features of life in order to explain the origin of life from matter, then it may be concluded that this program has failed. It then makes more sense to reverse the scientists' program and admit that since the cause of life must possess the properties of life, life itself must be the fundamental causative principle, and matter must be a derived phenomenon originating in life. This approach to nature is consistent with the ancient understanding, accepted by the seminal thinkers of both East and West, that the ultimate causative principle is the Supreme Living Being, and that all phenomena are derived from Him. It should also open up a wealth of new avenues of scientific exploration.

2

In Sections 1, 2, and 3 the basic arguments are developed. The implications of these arguments are described in Section 4. The mathematical definitions and derivations are included in the appendix.

1

**The laws of nature,
as understood by modern physics,
have a low information content.**

One of the fundamental principles of modern physics has been that the laws of nature can be described by very simple and general mathematical relationships. This is perhaps epitomized by the example of Albert Einstein, who strove during the major portion of his life to find a "unified field theory" which would derive all the forces and laws of interaction of the physics of his day from a single, simple unifying principle.

In physics, the laws of nature are generally studied only in highly restricted and simplified situations. This is done primarily for the practical reason that in more complex situations the mathematical analysis quickly becomes so difficult that no one can carry it out. The equations of quantum mechanics, for example, can be solved only for a system containing two particles at most. For larger systems approximations must be used, and these become very difficult to handle even in the case of the diatomic hydrogen molecule (which has four particles). For still larger systems, such as simple organic molecules, only guesses and conjectures can be made about the predictions of quantum mechanical theory. For large organic molecules such as those found in living cells, it is hopeless even to attempt to give a quantum mechanical description.

Nonetheless, the theories of physics are regarded as being, at least in principle, complete and universal descriptions of the phenomena of nature. It is believed that the reality of nature (or that which is really there) can be described by systems of numbers, and that all of the phenomena of nature can be calculated from certain mathematical relationships between numbers which are called "the laws of nature." Furthermore, it is accepted that these fundamental laws are simple in form and can be ascertained by human scientific investigation.

This assumption is very striking from the viewpoint of mathematics, where it is appreciated that theoretical models possessing any number of axioms, or basic laws, can exist. Indeed, it has been shown that any model of the natural number system will be logi-

4

cally ambiguous unless it contains an infinity of basic axiomatic laws. (This is Gödel's incompleteness theorem.) The assumption of the physicists that a very few basic laws will suffice to describe nature is therefore very restrictive, even though it is a practical necessity if the goal of physics—the determination of these laws—is to be realizable.

It is now an almost universal belief among scientists that this basic program of physics has been successfully carried out, at least for phenomena involving moderate masses, temperatures, and velocities. Specifically, it is believed that all of the phenomena of chemistry follow the known physical laws, and that all of the phenomena of life can be reduced to chemistry (and, hence, ultimately to physics). Thus the biochemist James Watson declares that with the development of quantum mechanics, "the various empirical laws about how chemical bonds are formed were put on a firm theoretical basis. It was realized that all chemical bonds, weak as well as strong, were based on electrostatic forces."[1] In his description of the basic goal of molecular biology, he states, "we see not only that the laws of chemistry are sufficient for understanding protein structure, but also that they are consistent with all known hereditary phenomena. Complete certainty now exists among essentially all biochemists that the other characteristics of living organisms (for example, ... the hearing and memory processes) will all be completely understood in terms of the coordinative interactions of small and large molecules."[2]

The known laws of quantum mechanics must indeed be very remarkable entities, since it is proposed that all the features and characteristics of life depend on these laws. Let us therefore try to write them down.

First of all, let us consider the laws of nature in classical physics. These can be summed up by the following equations:

$$\frac{dq_j}{dt} = \frac{\partial H(p_1, \ldots, p_n; q_1, \ldots, q_n)}{\partial p_j} \tag{1}$$

$$\frac{dp_j}{dt} = -\frac{\partial H(p_1, \ldots, p_n; q_1, \ldots, q_n)}{\partial q_j} \tag{2}$$

In classical physics, the state of a physical system at any given time is completely described by the position corrdinates, q_j, and

5

momentum coordinates, p_j. This is a set of 6N numbers for a system of N material particles, and equations (1) and (2) describe how they change with the passage of time. The function, H, which is called the Hamiltonian, is generally given by a simple formula in classical physics:

$$H = \sum_{j=1}^{n} p_j^2/(2m_j) + V(q_1, \ldots, q_n) \tag{3}$$

$$V = \sum_{i<j} A_{ij}/|\bar{r}_i - \bar{r}_j| \tag{4}$$

$$|\bar{r}_i - \bar{r}_j| = \sqrt{(q_{3i}-q_{3j})^2 + (q_{3i+1}-q_{3j+1})^2 + (q_{3i+2}-q_{3j+2})^2} \tag{5}$$

We have written out these formulas in full in order to show how very simple the laws of classical physics are. This is quite literally the full extent of the laws of nature as understood in classical physics up to the time of Maxwell. According to those who adhered purely to the scientific philosophy that nature could be completely described by mathematical laws, all of the phenomena of nature are consequences of equations (1) through (5) and the initial values of the q_j's and p_j's at some arbitrary starting time, t = 0.

This philosophy was given its initial impetus in the 18th century by Isaac Newton. He summed it up as follows: "I . . . suspect that [the phenomena of nature] may all depend upon certain forces by which the particles of bodies . . . are either mutually impelled towards one another and cohere in regular figures, or are repelled and recede from one another."[3] Equation (4) specifies these forces, which attract if A_{ij} is positive and repel if it is negative. The same view was expressed more recently in the 19th century by the physicist Hermann von Helmholtz: "The task of physical science is to reduce all phenomena of nature to forces of attraction and repulsion, the intensity of which is dependent only upon the mutual distance of material bodies. Only if this problem is solved are we sure that nature is conceivable."[4] Needless to say, these scientists include life as a "phenomenon of nature."

With the advent of Maxwell's electromagnetic theory, Einstein's theory of relativity, and the theory of quantum mechanics, the simple view of nature summed up by equations (1) through (5) underwent a considerable change. However, the basic sentiment expressed by Helmholtz that natural phenomena should be reducible to the interplay of elementary material forces has been retained. In the present dominant theory—quantum mechanics—physical systems are still described by arrangements of numbers, although the particle coordinates of classical physics have given way to Hilbert space vectors. The laws of transformation of these numbers are still given by brief equations which may be written down in a few lines.

In quantum mechanics, the basic equation of motion for a physical system is the Schrödinger equation:

$$i\hbar \frac{\partial \psi}{\partial t} = H\psi \tag{6}$$

Here, the state or exact physical description of the physical system is given by the Hilbert space vector, ψ, which can be represented in various alternative ways as a mathematical function or as a sequence of numbers. The Hamiltonian function, H, has been adopted from classical physics and now appears as an operator capable of acting on ψ to produce a new vector. In analogy with equation (3), H could be given by:

$$H = \sum_{j=1}^{n} \frac{-\hbar^2}{2m_j} \frac{\partial^2}{\partial q_j^2} + V(q_1, \ldots, q_n), \tag{7}$$

where V is the same as in equation (4).

Equations (6) and (7), along with (4) and (5) and an initial value for ψ at the time, $t = 0$, completely specify the quantum mechanical picture of a physical system of n/3 particles moving according to the attractive and repulsive forces given by V.

In further developments of the theory of quantum mechanics, things become somewhat more complicated. In addition to equation (4), various other terms are added to V to represent different kinds of forces believed to be acting in physical systems. These include terms for "spin" and electromagnetic interactions. Also, the basic form of H in equation (7) is sometimes modified in various ways, as happens in relativistic quantum mechanics and quantum field theory. It remains true, however. that the Hamiltonian for any system which is supposed to represent the fundamental laws of nature can be expressed by very brief formulas. When the abbreviated notations in these formulas are written out in full, as done above for the classical case, the resulting equations will be found to occupy a few lines at most.

$$\text{(a)} \quad H\Psi = i\hbar \frac{\partial}{\partial t} \Psi$$

$$\text{(b)} \quad H =$$

$$\sum_n \frac{-\hbar^2 c^2 \frac{\partial^2}{\partial q_n^2} + \eta_n^2 q_n^2}{2} + \sum_k \frac{-\hbar^2 \nabla_k^2}{2 m_k}$$

$$+ \sum_k \frac{i\hbar e_k}{m_k c} \bar{A}(\bar{Q}_k) \cdot \nabla_k + \sum_k \frac{e_k^2}{m_k c^2} |\bar{A}(\bar{Q}_k)|^2$$

$$- \sum_k \frac{e_k}{2 m_k c} \bar{\sigma}_k \cdot \nabla_k \cdot \bar{A}(\bar{Q}_k) + \sum_{ij} \frac{e_i e_j - G m_i m_j}{|\bar{Q}_i - \bar{Q}_j|}$$

$$\bar{A} = \sum_n q_n \bar{A}_n$$

Figure 1. The Laws of Physics Underlying Chemistry

This is true in particular of the physical model of chemical interactions referred to above by Watson as being sufficient for a complete understanding of life. The Hamiltonian for this model should include terms for electric forces, spin interactions, and electromagnetic interactions (plus gravity). This Hamiltonian is illustrated in figure 1.[5]

It is our thesis that a system of equations as brief and simple in form as these cannot possess sufficient power of discrimination to summon forth from a chaos of randomly distributed atomic particles the complex and variegated world of life we see about us. The theory of the origin of life from inanimate matter invokes two processes: chance and natural selection. The idea is that "chance" will provide various combinations of molecules which may or may not be useful in living organisms, and "natural selection" will pick out those which are useful and eliminate those which are not. Geneticists such as R. Fisher have argued statistically that even if natural selection only slightly favors one combination or form over another, in a sufficient length of time the favored form will nonetheless be found to predominate over the unfavored one.[6]

However, natural selection must have some direction if it is consistently to choose certain material configurations out of the myriads of configurations possible. The local selective advantages within particular populations must add together (as in a vector sum of vectors added tail to head) to a general trend from primordial soup to higher organisms. Ultimately, the fundamental laws of nature must provide this direction. At least, this must be true if nature is indeed to run in accordance with such laws.

It is very hard to see, however, why "forces of attraction and repulsion . . . dependent only upon the mutual distance of material bodies" should select trees, amoebas, bumblebees, or human beings in favor of other possible material configurations, such as inert globs or blotches. Enhancing the theoretical picture with spin interactions following the Pauli matrices, or electromagnetic fields composed of harmonic oscillators, does not seem to add more plausibility to the idea that natural selection could do this. We will argue that the very brevity of the laws of nature as they are expressed in physical theories makes them unsuitable for selecting the complex forms of living organisms from an initial state of molecular chaos, no matter how much time is allowed for the process. Basically, we shall show that in order for a set of natural laws to select a complex form out of a random distribution of matter, these laws must possess a corresponding level of complexity themselves. This will imply that the Hamiltonian for a system in the quantum mechanical formulation would require many pages of symbols to write down (as many as fifty at the very least) in order for that system to select configurations with the complexity of living organisms. In other words, the known laws of physics are insufficient to account for the origin of life, and in order for a system of physical laws to do so, its sheer size and complexity would make it impossible for the human mind to handle. (We should point out that high complexity will be shown to be a necessary but not sufficient requirement for a system to exhibit the evolution of complex forms.)

As we have already pointed out, the equations of motion cannot be solved exactly for systems containing more than two particles in either the classical or the quantum mechanical theories of physics. Indeed, in the theory of relativistic fields these equations have given rise to difficulties even in the case of zero particles (the so-called vacuum state). Since systems capable of describing living organisms must contain enormous numbers of particles on the order of 10^{23}, we can see that it can never be practical to study

9

the nature of such organisms by explicitly solving the equations of motion. However, such solutions must exist in principle in order for the theory to be valid at all. The procedure of scientists, then, is to establish the existence of solutions by abstract reasoning and to demonstrate that these solutions could be calculated in principle. Then they attempt to apply and verify the theory by making logical deductions about the properties of the solutions without actually seeing them. We shall proceed on the assumption that the required solutions always do exist and could be calculated.

We shall formally measure the complexity or information content of a theory as the length of the shortest computer program which can numerically solve the equations of motion for the theory to within any desired degree of accuracy. The Schrödinger equation (equation (6)) can be solved in principle by a simple numerical algorithm. Consequently, the information content of a theory having this equation as its basic equation of motion is nearly proportional to the number of symbols needed to write out the Hamiltonian for that theory in terms of the specific programming language. For consistency, all estimates of information content will be referred to a fixed programming language. The information content of a configuration of matter, such as the body of a living entity, can also be estimated as the length of the shortest program which will generate a complete numerical description of the configuration. We shall use this measure of information content to provide a clearcut numerical demonstration that the known laws of physics, or any system of laws of a similar nature, should fail, even in principle, to account for the origin of life.

In a mathematical model of a physical system, two other ingredients are needed along with the laws of motion. These are the initial conditions and boundary conditions for the system. Normally, a physical system will be confined within a certain fixed volume of space: in order to describe the events within the system, it is necessary to describe the physical conditions along the boundary of this volume during the time in which the system is being studied. Also, in most physical models the system is considered during a finite time interval, $0 \leqslant t \leqslant t_1$. It is therefore necessary to specify the physical state of the system at the beginning of this time period, $t = 0$.

In a theory of the origin of life, the initial conditions should describe a "primoridal" situation possessing a very low degree of organization, if any. For example, most theories of the chemical origin of life postulate that life arose from a "primordial soup" consisting of a mixture of water and simple compounds such as CO_2, CH_4, N_2, NH_3 and H_2S and a reducing atmosphere composed mainly of CH_4 and NH_3.[7] This mixture of chemicals is presumed to receive radiation from the sun, to receive supplies of gases from the earth (volcanic venting), and to radiate heat and light into outer space. This sums up the initial and boundary conditions for this model.

As another model, one could start with the supposed origin of the solar system from a cloud of gas. The initial conditions would then consist of a description of the initial gas cloud, and the boundary would correspond to an unlimited vacuum surrounding this cloud (if we ignore the influence of distant stars). Then, according to theory, the laws of nature would presumably first generate the solar system, complete with primordial soup, and then generate life in the soup. Or, one might consider a model of the universe as a whole, such as the "big bang" theory, which features a superhot soup of subatomic particles as its initial condition.

In any case, the idea behind all theories of the origin of life from matter is that one only has to propose a simple set of conditions to hold in the beginning. After all, the idea of the theory is to "explain" all the features of life, and the greater the intricacy of the specifications required for the initial conditions, the less complete the explanation becomes. One then has to explain the origin of the intricate initial conditions in terms of some still earlier state of affairs.

A typical initial condition for a model of the origin of life will consist of an ensemble of possible initial states, such as one of the ensembles of statistical mechanics or some simple combination of these. This certainly holds true for the models referred to above. Such an ensemble can be specified by a simple equation or a brief set of equations. For example, in the theory of quantum mechanics one of the standard thermodynamic ensembles, known as the canonical ensemble, is given by the equation,

$$\rho_0 = K^{-1} \exp(-H/kt) \tag{8}$$

where H is a Hamiltonian operator.[8]

In this equation, ρ_0 is called a density matrix. It corresponds to a collection or ensemble of quantum mechanical states, each of which has a statistical weight assigned to it which gives the probability that that particular state will be found to hold in the physical system. This means that the initial state of the system is left ambiguous or undefined to a very large extent. In fact, one of the basic principles of statistical mechanics is the assignment of equal probabilities to all initial states satisfying certain simple restrictions, such as a certain particle density or a certain range of energies. (This is how the micro-canonical ensemble is defined.) In this case, the state of the system is completely ambiguous, apart from the requirement that these restrictions should be satisfied.

The idea of the theory of evolution is that out of this chaos or ambiguity, the laws of nature should be able to "select" molecular configurations capable of exhibiting all the phenomena of life. Another way of looking at this is the following: The laws of nature should have the ability to generate life forms from most of the possible initial physical states which satisfy certain simple physical requirements. This would assure that if one of these states were chosen "at random" as the initial condition of the system, life forms would probably appear in the system at some later time.

The boundary conditions for the physical model should also be definable in simple terms. At most, one would expect energy or material particles in a simple form to pass back and forth across the boundary. One can define these conditions by specifying external electromagnetic fields, and also probability distributions for incoming matter, such as cosmic rays. This can also be done by means of simple equations. In some models the boundary might consist of reflecting or non-interactive walls, or a limitless void, or there may be no need for explicit boundary conditions at all. Basically, any interaction between the system and matter or energy beyond its boundary should be describable in simple statistical terms, as can be done, for example, for cosmic rays or the influx of solar radiation. Boundary conditions, like initial conditions, must be relatively simple: if intricate specifications were required for the boundary conditions, one would need to explain their origin also.

Once the initial and boundary conditions are defined, the state of the physical system at each subsequent time, t, between 0 and t_1 must be determined by the natural laws of the system. The basic situation is summed up in figure 2. A theory of the origin of life from matter should require that in some range of times, perhaps in the range of three or four billion years or so, these states will exhibit the molecular configurations characteristic of living entities with a reasonably high probability. Let us consider how

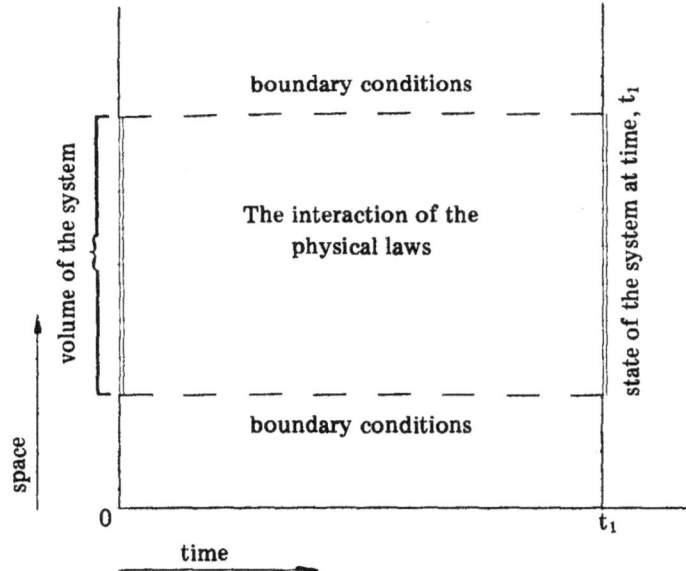

Figure 2. The basic form of a physical model

these configurations may be mathematically represented in the physical states. We will do this for the theory of quantum mechanics, in particular. In the theory of quantum mechanics, any entity or material configuration one might want to observe in the system is represented by a mathematical operator called an "observable." Let us donate the entity to be observed as B and denote the observable to which it corresponds by B. Then the probability of finding B in the system represented by the density matrix, ρ, is given by,

$$\text{Prob (B)} = \text{Trace } (\rho B) \qquad\qquad (9)$$

All that we want to point out about this formula is that it can also be calculated by a brief set of computer instructions.

We are interested in the case where B is a particular configuration of atoms representing a molecule or collection of molecules. One way of describing such a configuration is by means of a numerical code indicating which atoms should be found in close proximity to which other atoms—that is, indicating the pattern of chemical bonds which characterizes the configuration. If X repre-

13

sents such a configuration, then the observable, B(X), corresponding to it can be expressed by a simple formula (see the appendix).

Thus, the boundary conditions, initial conditions, and laws of nature determine the density matrix, ρ, describing the physical state of affairs at time, t_1. From ρ we can (at least in principle) calculate the probability, M(X), that any particular configuration described by a code, X, can be found somewhere in the system at time, t_1 :

$$M(X) = \text{Trace } [\rho B(X)] \tag{10}$$

The probability function, M(X), determines what configurations can be expected to have evolved in the system by the time, t_1. If M(X) is very large, X can be expected to have evolved, but if it is very small, the evolution of X is unlikely. If we measure the information content of a number or a mathematical function as the length of the shortest computer program which will calculate it, then it follows that the information content, L(M), of the function, M, is no greater than the total length of all the various calculations we have just described. This should correspond to about three or four pages of tightly packed programming instructions at the most. (We envision this program as being written in some standard programming language with provisions for handling numbers with an arbitrary number of significant digits.)

We shall show in a later section that M(X) must be exceedingly small if the information content, L(X), of X is very much greater than the information content, L(M), of M plus a certain fixed constant. The difference between the information in X and that supplied by the system (represented by M) must be made up by pure chance, and therefore the probability of X goes down exponentially with this difference. This shows that the evolution of life forms in such a system is exceedingly unlikely, for it stands to reason that the length of the shortest program needed to calculate the essential molecular structures of even a "primitive" living organism should be a great deal longer than three or four pages. This is argued in detail in the next section.

We should note the ways in which "chance" enters into the expression, M(X), for the probability of finding X. In both the initial conditions and the boundary conditions, there is randomness due to the thermodynamic ensembles expressing them. This is the source of the randomness postulated for the "random mutations" of the theory of evolution, which are supposed to be due to chance molecular collisions and collisions with cosmic rays. Also, randomness is built into the basic structure of quantum mechanics, since the quantum mechanical states are statistical distributions. Both the mutations and the natural selection of the theory of evolution are built into these physical models, as we would expect.

As a final point, we would like to mention one other model that exhibits all of the features involved in the theory of evolution. This is the cellular automaton model of John von Neumann. In this model, we suppose that a small automaton, which can be in any one of a finite number of states, is situated on each square of a large two-dimensional lattice. The state of the system is given by specifying the states of each small automaton in the lattice. The system changes with time in the following way: Let t be a small, fixed time interval. At the end of each successive time interval, t, each automaton changes to a new state in a way which depends only on the states of the automatons on the squares immediately adjacent to it. The transition function which determines these changes can be specified by a few lines of instructions.[9]

Von Neumann has shown that a self-reproducing machine capable of performing complex calculations (a universal Turing machine) can be constructed in this system in the form of an arrangement of a large number of these small automatons. His idea was to prove that the property of self-reproduction characteristic of living organisms can be exhibited by a mechanical system. This was intended as one further step in demonstrating that life is a mechanical process governed by mathematical laws.

It is therefore interesting to inquire whether self-reproducing systems of the sort which von Neumann considered could evolve in his cellular automaton system in a sufficient length of time. The law of transformation of the system with time should determine the natural selection of different forms in preference to others within this system by the same sort of processes of competition envisioned in ordinary evolutionary theory. The random mutations of evolutionary theory could be easily introduced in the form of a Markov process which would make random changes in the states of the small automatons.

However, such evolution could not be expected to occur. Owing to the great simplicity of von Neumann's system, the function corresponding to M(X) for the system can be expressed by a very brief list of equations. Von Neumann's self-reproducing machines, on the other hand, are very complicated, and he re-

15

quired several hundred pages to describe them in his book. It therefore seems reasonable to suppose that their information content greatly exceeds that of M. As we shall see in Section 3, this rules out the evolution of such configurations within the system.

2

The bodily structures of living organisms have a high information content.

We shall consider the bodily structures of living organisms in three aspects: the visible bodily form (phenotype), the genetic DNA code (genotype) and other molecular structures of cells, and the behavioral patterns of organisms. Under the last heading, we shall be particularly concerned with human behavior and its by-products, such as language, literature, technology, and scientific theories.

First, let us consider the molecular structures of cells. One of the most thoroughly studied organisms is the bacterium *Escherichia coli*, a unicellular organism about 500 times smaller than the average cells of higher plants and animals.[10] It is one of the smallest and simplest of all living organisms. Yet it is estimated that a single *E. coli* cell contains between 3,000 and 6,000 different types of molecules. Among these are some 2,000 to 3,000 different kinds of proteins with an average molecular weight of 40,000, as well as a single DNA molecule with a molecular weight of 2.5×10^9. This molecule of DNA is believed to contain coded instructions for the construction of all the other molecules. James Watson, one of the foremost authorities on molecular biology, admits that these molecules do not obey any simple structural rule:

> Most of these macromolecules are not being actively studied, since their overwhelming complexity has forced chemists to concentrate on relatively few of them. Thus we must immediately admit that the structure of a cell will never be understood in the same way as that of water or glucose molecules. Not only will the exact structures of most macromolecules remain unsolved, but their relative locations within cells can be only vaguely known.[11]

It is believed that protein molecules may be described as chains of amino acid molecules, which may be of twenty different varieties. A typical protein molecule in an *E. coli* cell will contain some 300 of these amino acid subunits. Since each subunit may be any one of twenty different amino acids, this means that the number of possible protein molecules of typical size is about 20^{300}. The information content of a typical protein molecule in an *E. coli* cell is therefore bounded by an upper limit of L(X) = 300 $\log_2$ 20 $\simeq$ 1290 bits. Since there are an estimated 2,000 to 3,000 different proteins of this kind in a cell, the total information content for cellular protein is bounded by an upper limit of between 2,580,000 and 3,870,000 bits.

The DNA molecules of cells are thought to be helical chains composed of successive pairs of DNA bases. There are four different kinds of bases—adenine (A), thymine (T), guanine (G), and cytosine (C)—and these are limited to forming four different kinds of base pairs: A-T, T-A, C-G, and G-C. The DNA molecules within a cell are believed to contain coded instructions for the structure of each protein molecule within the cell. Each group of three base pairs along the DNA chain specifies a specific amino acid in a corresponding protein molecule, or it codes for the termination of a protein chain. Each DNA base pair has a molecular weight of about 660, and so a group of three pairs has a molecular weight of about 2,000. Since there is a molecular weight of about 2.5×10^9 for the (haploid) *E. coli* DNA, this means that the genetic code for this organism will consist of some 1.2×10^6 triplets of bases. Since each triplet can discriminate between 21 alternatives (20 amino acid types and a stop code), this gives us an upper bound of about 1.2×10^6 times $\log_2 21$, or some 5.2×10^6 bits, for the genetic information content of an *E. coli* cell.

Yet the *E. coli* is a very simple cell. In the cells of higher plants and animals, much larger amounts of DNA are found than in *E. coli*. It is estimated that mammalian cells contain some 800 times as much DNA, yielding an upper limit of some 4.2×10^9 bits for the genetic information content of these cells.[12] Some idea of the size of these numbers can be obtained by considering the number of pages of print that would be required to write down these amounts of genetic coding in full. In a typical book there are some 70 characters per line and 40 lines per page. This gives us about 1.7×10^4 bits per page if we use an alphabet of 64 characters. (Since 64 = 2^6, we can code 6 bits per character.) At this rate it would take about 300 pages to write down the coding for *E. coli*, and about 240,000 pages to write down the coding for a mammalian cell.

One of the dogmas of modern molecular biology (called the central dogma, in fact) is that all the information needed to specify a cell is contained in the cell's DNA coding, and that this coded information is not changed except by random mutations. Whether this is true or not, the amount of DNA in cells should give some idea of the amount of information they contain. According to the central dogma, the figures we have derived thus far should provide upper limits for cell information content. We would now like to estimate reasonable lower limits. To do this, we shall need to consider the variety and complexity of the structures cells are found to contain. Since some genes in the chromosomes of higher animals apparently exist in multiple copies, it would seem that their total genetic information content should be lower than the figures suggested by their total DNA content. However, based on the great complexity of the structures that can be seen in vertebrate cells under the microscope, Watson estimates that such cells should be "at least 20 to 50 times more complex genetically than *E. coli.*"[13] By this he means that such cells should contain coding for at least 20 to 50 times as many protein macromolecules as *E. coli.* We would need between 6,000 to 15,000 pages to write down this amount of coding, based on our 300-page estimate for *E. coli.* This corresponds to between 1.0 and 2.6×10^8 bits of information, and between 80,000 and 200,000 proteins, with an average of 300 amino acid subunits apiece.

In order to determine a lower limit for information content, let us consider how the genetic information is to be distributed along the length of the chromosomes in a cell of a higher plant or animal. The total DNA coding of such a cell should be representable by a binary number of 4.2×10^9 digits, or, using an alphabet of 64 characters, by a character string of 7×10^8 characters. Let us call this character string X. We can divide X into N blocks, $Y_1, \ldots, Y_N$, each of which has a length of $M = 7 \times 10^8 /N$ characters. Call the string made up of the first k of these blocks (going from left to right) X_k, for each integer, $k = 1, \ldots, N$.

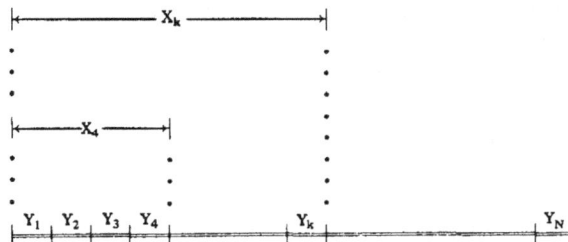

Figure 3. Division of the genetic coding sequence, X, into a series of adjacent blocks.

The following inequality relates the total information content of X with the incremental amounts of information required to express each successive Y_k in terms of X_{k-1}.

$$L(X) \geqslant L(Y_1|I) + L(Y_2|X_1,I) + \ldots + L(Y_k|X_{k-1},I)$$
$$+ \ldots + L(Y_N|X_{N-1},I) - Nc \qquad (11)$$

The term, $L(Y_k|X_{k-1},I)$, is the amount of information needed to calculate the string, Y_k, given that we already know the string, X_{k-1}. (Note that $X_1 = Y_1$ and $X = X_N$.) Specifically, this term is the length in characters of the shortest program that will calculate Y_k using X_{k-1} as a possible source of data. The term, $L(X)$, is the total information content of the string, X, in characters. The constant, c, is an artifact of the mathematical derivation of this inequality, and is no more than about 47 characters.

Inequality (11) can be rephrased as follows:

$$\frac{L(X)}{7 \times 10^8} \geqslant \frac{\text{Average of } L(Y_k|X_{k-1},I) - 47}{M} \qquad (12)$$

where the average is over $k = 1, \ldots, N$.

Suppose, for the sake of argument, that $L(X) \leqslant 10^5$ characters. Then, if we set M, the length of the blocks, Y_k, equal to 7,000, we find that on the average, $L(Y_k|X_{k-1},I) \leqslant 48$ for $k = 1, \ldots, N$. ($N = 10^5$) This means that a typical block, Y_k, of 7,000 characters can be calculated exactly from X_{k-1} by means of a program no more than 48 characters in length.

Let us consider what this means. We have postulated that the information content of the genetic coding string, X, is only 1/7,000 of its maximum possible value. This implies that, on the average, each successive block of 7,000 characters along X can be exactly calculated from the previous blocks by a computer program of no more than 48 characters in length. Now, a block of 7,000 characters corresponds to several genes specifying protein macromolecules. For example, nearly 33 genes specifying proteins of 300 amino acid subunits in length can be coded with 7,000 characters. Since the amino acid sequences of such proteins appear very complicated, it is hard to believe that they could be consistently transformed from one to another by such short programs.

These proteins play very sophisticated roles in the metabolism of cells. The exact way they function is far from known at the present time. However, it is known that they are able to behave like small computers. Here is an example taken almost at random from Watson:[14] In *E. coli* there occurs a sequence of chemical reactions that convert the compound threonine into isoleucine in five steps. The first step in this sequence can occur only with the aid of the enzyme L-threonine deaminase, a protein macro-molecule. When the final product, isoleucine, has reached suf-ficiently high concentrations, it interacts with the enzyme molecules in such a way that they no longer catalyze the first reaction of the series. This prevents the manufacture of more of the product than needed in the cell. We may note that the enzyme is so structured that it catalyzes only one specific reaction. While not affecting the rates of other chemical reactions at all, such biological enzymes are known for their ability to cause certain chemical reactions to occur millions of times faster than they will occur under laboratory conditions in which the enzyme is not present. We may also note that this enzyme is inhibited specif-ically by isoleucine at the proper level of concentration, and not by any other chemical that would normally be present in the cell (for otherwise this control system wouldn't work.) Many enzymes of this sort, which interact with other molecules according to a logical scheme and perform very specific chemical operations in a highly efficient way, are evidently required if the cell is to func-tion as a highly versatile and complex automaton. It would there-fore seem very remarkable if the typical block of 33 such enzymes coded on X could be exactly calculated from the previous blocks in the chain (namely, X_k) by a program of only 48 characters. This would be like expecting the first few hundred decimal digits of π or $\sqrt{2}$ to code for an *E. coli* enzyme.

Yet we have estimated that there may be as many as 200,000 such enzymes in a mammalian cell. A block of genes could be simply computed from the previous blocks in the chain if it were a duplicate copy or very near copy of the genes in the previous blocks. But we would not expect this to be possible very often for genes performing functions distinct from those in the previous blocks. We conclude that we must be incorrect in our assumption that the total complexity, L(X), of the genetic code for a mam-malian cell is less than 10^5 characters. We therefore propose that this figure can be taken as a reasonable lower bound for L(X).

Lower bound of $L(X) > 10^5$ characters.   (13)

21

This lower bound corresponds to 1/7,000 of the upper bound of L(X). Watson[15] indicates that some of the genes (but, presumably, by no means all) in cells of higher animals may appear in as many as 100 to 1,000 redundant copies, but this could account for only about 1/70 of the redundancy our lower bound allows. Therefore it might also be reasonable to consider a lower bound of 7×10^6 for L(X). As we have noted, Watson proposes that the cells of higher animals must have some 20 to 50 times as many distinct protein macromolecules as *E. coli.* If these proteins can be specified in a chain only by programs with an average length of no less than 57 characters, then the minimal value of L(X) will be between 8×10^5 and 2×10^6. [This follows from inequality (11) with $57 \leqslant$ Average of $L(Y_k|X_{k-1},I)$ and 80,000 to 200,000 for N. In this case we let X represent the sequence of codes for the different proteins.]

In order to give a further indication of the type of complexity we are dealing with in living cells, let us now describe some of the different categories of structure that organisms display. We have ruled out the possibility that very short programs can transform the sequence of genes in X_{k-1} into the sequence in Y_k. Actually, for the overwhelming majority of sequences of characters,

$$L(Y|X,I) \simeq L(Y|I) \simeq L(Y) \simeq \text{length of Y.} \tag{14}$$

For proteins of 300 amino acid units in length, we have length of $Y = 300 \log_{64} 20 \simeq 215$ characters. A very elementary theorem about the information function, L, indicates that no more than $1/64^{(215-57)}$ of the total number of Y's of 215 characters can be calculated from a given X by programs of 57 characters or less. Almost all sequences of symbols of a given length can be specified most briefly by simply writing them out in full. Of those that remain, the overwhelming majority can be most simply specified by writing them out nearly in full and indicating some brief calculations that will compute the remainder of the sequence.

In view of this, consider the following hierarchy of structures exhibited by living organisms.

(a) *The chemical reactions involved in cellular metabolism.* These involve respiration, the synthesis of various chemicals needed in the cell from food molecules, photosynthesis in plants, and the processes involved in the orderly breakdown of different molecules. It would appear that most of the genetic coding of *E. coli* must be devoted to this, since these bacterial cells do very little but grow and divide in half. Even though *E. coli* is one of the simplest of organisms, its metabolic interactions are so intricate that "the exact way in which all these transformations . . . occur is enormously complex, and most biochemists concern themselves with studying (or even knowing about!) only a small fraction of the total interactions."[16] That these interactions must be governed by a complex system of logic rivaling the most sophisticated programs of modern electronic computers is certainly indicated by the descriptions given in Watson's book.

(b) *The morphological structures of single cells.* The *E. coli* cell appears to possess a relatively simple gross structure, but many cells, even among the bacteria and protozoa, possess very intricate structures. For example, the cilia of protozoa such as the paramecium have been shown to possess a very intricate structure that elegantly exploits certain mechanical principles to produce a synchronized rowing machine.[17] Even the *E. coli* cell possesses some remarkable mechanical features. This bacterium propels itself through the water by means of a spiral flagellum that is spun about its axis by a motor built into the bacterial cell wall.[18] This motor is complete with drive shaft and some kind of rotating disks, but the principle underlying its operation is still unknown. It seems very doubtful that structures such as these can be calculated by computer programs of 57 or so characters, or that they can be transformed by such programs into, say, the coding for the enzymes involved in the Krebs cycle of cellular respiration. (Note that the preceding sentence is 225 characters in length.)

(c) *The different types of cells involved in the organs of multicellular organisms, and their different functions.* We can easily write down a long list of different types of cells appearing in different bodily organs. These include muscle cells, nerve cells, bone cells, different kinds of blood cells, glandular cells, liver cells, epithelial cells, etc. The study of a particular kind of cell can be a whole academic subject in itself, and doctoral dissertations are frequently devoted to the study of a detail of a detail of the structure and function of such cells. The complete instructions for constructing all these different cells must be contained in the genetic coding of any higher organism, at least according to the understanding of modern biology. There must also be instructions controlling the development of these various types of cells during the growth of the embryo.

(d) *The structure and function of different types of organs in higher plants and animals.* This is an aspect of (c). The different

23

organs of the body perform a vast array of complicated functions, most of which are either unknown or poorly understood. Examples include the disease fighting system of the blood, the image-producing eye and its retina, the brain, the endocrine gland system, and the heart and circulatory system. It also seems very doubtful that the coding for different features of these various organs could be transformed into the coding for the exact specifications of other organs by computer programs of a few characters in length.

As two contrasting examples of such organs, let us consider the eye and the feather of a bird. Both of these examples have long troubled the more thoughtful students of the theory of evolution because it is very hard to see how they could have originated by natural selection and mutation. Thus, Charles Darwin said, "I remember well the time when the thought of the eye made me cold all over, but I have got over this stage of the complaint, and now small trifling particulars of structure often make me very uncomfortable. The sight of a feather in a peacock's tail, whenever I gaze at it, makes me sick!"[19]

A brief consideration of the structures making up the eye will give some indication of why Darwin felt as he did. These include the lens, the muscles supporting it, the iris, the cornea, the retina, the nerve connections, and the muscle system that moves the eyeball. Many of these structures are very intricate. For example, the iris contains a muscle system for opening and closing the pupil; the lens must be transparent and shaped so as to focus a sharp image on the retina; the retina contains systems of cells and nerves designed to detect different elementary visual patterns such as lines and edges; the light-sensitive cells contain complex chemical systems designed to respond to different colors of light; and so forth. All these structures must be coded into the genetic specifications of the organism, and it is hard to see how the codes for such diverse structures could be transformed, one into another, by simple computer programs. On the contrary, it seems much more reasonable to suppose that the shortest set of instructions defining the eye should contain detailed descriptions of each of these structures individually. In other words, the eye should have a high information content.

The feather is also a very complicated structure. Alfred Russell Wallace, the co-founder with Darwin of the natural selection theory, later rejected this theory because he felt that it could account neither for the nature of man nor for the amazing structures of living organisms in the plant and animal kingdoms. One of his examples is the bird's feather, which, as he points out in his book *The World of Life*. contains many different structural features of great intricacy.[20] We may also note that, superimposed on this basic structure, many feathers display very precise and detailed colored designs. The peacock feather, which so troubled Darwin, contains a multicolored image of an eye, set off from its background by a systematic thickening and thinning of the smaller strands of feather that make it up. These structures should also have a high information content.

(e) *The behavior of animals other than man.* Many different patterns of complex behavior are exhibited by lower animals. We may note, for example, the social systems of bees and ants, the spinning of spider webs, and the transcontinental migrations of birds. These different behavioral patterns are thought to be built into the genetic material of the organisms, and, as such, they must be coded in terms of a series of logical "if-then" instructions equivalent to the program of a computer. It might be an interesting challenge for a student of animal behavior to try to write computer programs that would duplicate the behavior of a particular animal. As a clue for the difficulty this would involve, consider the problem of pattern recognition. Many types of birds are highly discriminating as to the coloring and physical shape of other birds with which they relate in the course of their life. This means that they are able to make fine distinctions between various complex patterns of color and form. However, it has proven very difficult to write computer programs that can discriminate between printed letters.

We can therefore see that the coded instructions that specify animal behavior may be expected to be very complex. Also, we can see no particular reason to suspect that these instructions can be specified by a simple transformation of the instructions for any of the structures or systems of structures in categories (a) through (d). Rather, we should expect that these different categories of information should be independent, in the sense that knowledge of one of them, say the code for the antibody system, would not provide much knowledge about another, such as the coding for transcontinental migration. In symbols this means that we should expect $L(X) \simeq L(X|Y)$ and $L(Y) \simeq L(Y|X)$ if X and Y are the two different instruction codes.

(f) *Human behavior.* This is perhaps the most complex topic of all, and is at present quite beyond the reach of the reductionistic speculations of modern science. Most features of human behavior are believed to be independent of the genetic coding system of heredity and are thought to be transmitted from generation to generation by cultural learning processes. The evolution of human behavior is therefore regarded as something different from organic evolution and is called "cultural evolution," although it is presumed to occur according to the same laws as organic evolution. Indeed, modern science recognizes no laws other than the blind mathematical laws of molecular interaction.

Since human culture is an aspect of life, we may legitimately consider it in our estimate of the complexity or information content of living systems. From the reductionistic view of man as a system of molecules, it follows that the information of human culture must be coded in different bodies in some sort of molecular patterns, whether DNA or not. We may gain some idea of the amount of information involved in human culture by considering the numbers of volumes of books to be found in different libraries. To be sure, there is a great deal of redundancy in these books. We wonder, however, whether the total information content of the brain of a single learned person could be as low as our minimum estimate of the information content of a higher living system: 10^5 characters, or about 36 pages.

In order to rule out the bizarre thought that the information content of human knowledge could be this low, we can again perform the breakdown of total information into coded segments as described in inequality (11). If X now represents the text of a series of works of human literature, involving such fields as science, music, and philosophy, we could regard Y_k as a particular passage from one of these texts. If $L(X)$ is very low, then $L(Y_k|X_{k-1},I)$ will have to be very low on the average, just as in our previous analysis. This would mean that Y_k can be computed from X_{k-1} by a very short program of, say, 48 characters or so. That *Principia Mathematica* can be transformed in this way into the first act of *Macbeth*, or that these together can be thus transformed into the *Eroica* symphony seems very doubtful, to say the least. We therefore conclude that the information content of human culture must be a great deal higher than 10^5 characters.

As one final point, consider again the breakdown of the total genetic code sequence into segments, $Y_1, \ldots, Y_N$. This time, let the lengths of the segments be 700 characters, so that each Y_k specifies about three genes. As before, X_k represents the sequence formed by joining together $Y_1, Y_2, \ldots, Y_k$. Then there is *one particular program*, G_w, with an index $w = 1,2,3, \ldots$, which will transform each X_k into Y_{k+1}.[21] That is, for each k between 1 and N there is a number, w_k, for which

$$Y_{k+1} = G_{w_k} (X_k). \tag{15}$$

This program can be specified very simply in one line.

Now, if $L(X) \leqslant 10^5$, it turns out that for over 90% of the segments, Y_k, we have

$$Y_{k+1} = G_1 (X_k). \tag{16}$$

This implies that the simple transformation, G_1, serves as a magic formula which by itself is able to generate most of the genetic coding of a human being by repeated application. For most of the remaining segments, G_2 or G_3 will transform X_k into Y_{k+1}.

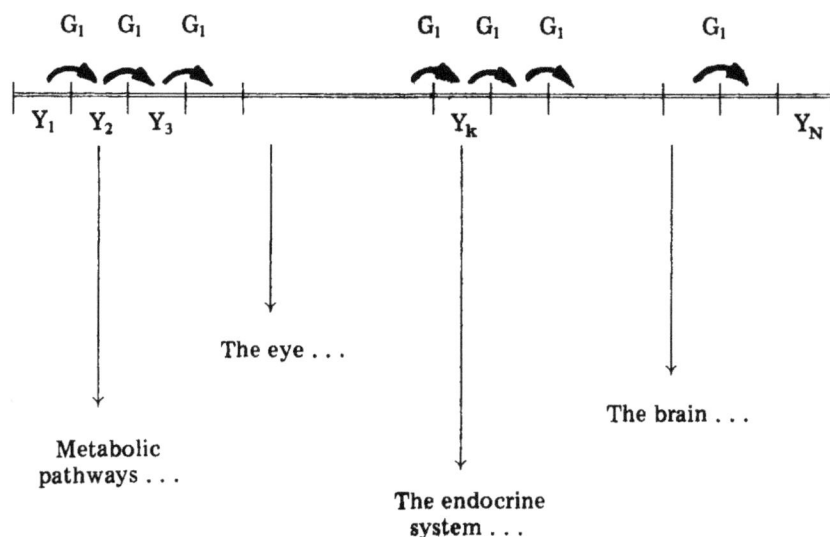

Figure 4. The magic formula, G_1. Can repeated applications of this simple one-line program generate large stretches (over 90%) of the genetic code? This is implied if this code has a low information content. $[L(X) \leqslant 10^5$ char.]

27

If we consider the great variety of information that must be encoded in X, it seems very implausable indeed that great stretches of this information could be simply reeled off by repeatedly applying the transformation, G_1. This would mean that successive applications of G_1 would generate coding for eyes, hands, the brain, the liver, metabolic pathways, and so on, as we have already described. It seems far more reasonable to suppose that $L(X)$ cannot in fact be as low as 10^5 characters.

3

No system of high information content can evolve from a system of low information content.

In section 1, we demonstrated that the models of modern physics for the course of events within the universe have a low information content. Our basic format for a physical model is as follows: We have a system, S, in which events take place in accordance with a set of natural laws, F. Generally F will consist of a differential equation or set of differential equations. S will have initial conditions, S_0, and boundary conditions, B. These will generally be described by probability distributions, s_0, and b, which are defined by ensembles of statistical physics such as the Gibbsian canonical ensemble. The probability distribution, s_t, for the system, S, at the time, t, can be calculated, at least in principle, from the equations for s, b, and F by a short computer program. By "short" we mean that this program can be written down in full using no more than four or five pages of solid print (at 70 characters/line and 40 lines/page.)

From the probability distribution, s_t, we can calculate the probability, M(X), that a particular molecular configuration described by a code, X, will be found somewhere within the system, S, at the time, t. M(X) is the probability that the configuration, X, will have "evolved" within the system by the time, t. Given X, the value of M(X) can be calculated from s_t by a few more lines of programming instructions. Thus, the total information content of the function, M(X), i.e. the value of L(M), can be estimated to be no more than five or six pages of computer instructions. This corresponds to no more than 14,000 to 16,800 characters of information:

Upper bound for L(M) $\leqslant$ 16,800 characters.   (17)

On the other hand, in Section 2, we estimated that the lower bound for the configuration, X, describing the genetic information for the cell of a higher animal could be no less than:

Lower Bound for $L(X) \geqslant 100,000$ characters.   (18)

The implications of this for the theory of evolution are given by the following inequality:

$$M(X) \leqslant 64^{c + \log T + L(M) - L(X)},   (19)$$

where c is a constant of 32 characters and T is given by:

T = The maximum number of distinct configurations, X, that
     can appear within one given state of the physical   (20)
     system, S.

This T is no greater than the volume of S divided by the minimum volume within S occupied by a configuration, X. For the volume of S let us take 10^{12} cubic kilometers as an upper bound, since this is greater than the volume of the Earth plus atmosphere. Since we are interested in molecular configurations, let us take one cubic Ångström unit as the minimum volume of a configuration. This gives us $T = 10^{51}$, so that $\log T \simeq 28$ using the base of 64.

$$M(X) \leqslant 64^{c + 28 + 16,800 - 100,000} \ll 64^{-80,000}   (21)$$

Inequality (21) says that the probability that the configuration, X, will be found anywhere within the volume of the system, S, at the time, t, can be no greater than one out of $64^{80,000}$. Suppose that t is 4.5 billion years—the estimated age of the Earth according to present geological theories. Multiplying M(X) in inequality (21) by $365 \times 4.5 \times 10^9$, we can see that the probability of finding X in the system at the end of any day in a 4.5-billion-year period is also much less than $64^{-80,000}$. We conclude that the entire course of events in S covering a 4.5-billion-year period would have to be repeated over and over again at least $64^{80,000}$ times in order for there to be a reasonable expectation that the configuration X would be seen in S even once. (This is in accordance with what is known as the frequency interpretation of probabilities.) Since $64^{80,000}$ is an enormous number, to say the least, it is reasonable to say that X will not evolve within the system S in 4.5 billion years.

Now, X could represent the genetic code for a higher animal, and S could be a physical model of an earth sized physical system which might be considered as a candidate model for the evolution of life. Our conclusion is that no higher animal will ever evolve in such a system in any realistic span of time. For our time, t, we could have chosen any time period for which L(t) is small, for the only property of t that affects our considerations is the number of characters needed to express t in the program for calculating the events in S. This means that our t could be anything from one year to billions of billions of years and the conclusion expressed in inequality (21) would still hold.

It is often said that while it is highly unlikely that any given complicated structural part of a living organism will arise by chance, still, given the immense spans of geological time, such structures are bound to arise sooner or later. When they do arise they will be preserved and propagated by the forces of natural selection, and in this way evolution will occur, even though it may seem improbable. The analysis presented here proves that this actually cannot happen in a system governed by simple natural laws and possessing simple initial and boundary conditions. The reason for this is easy to understand. The process of natural selection can only select simple configurations if the basic physical laws on which it runs are simple. Thus, even though various sequences of random change may occur over vast spans of time, simple processes of physical interaction will only be able to select simple patterns from among them, no matter how long the time span.

For this reason, the theory of evolution can be seen to fail at the very weak point that has been criticized by so many students of the theory. Many observers have noted that "natural selection" has never been adequately defined in evolutionary theory, either in Darwin's original version or in the more recent "synthetic theory." According to Darwin, natural selection means "survival of the fittest," but no one, unfortunately, has been able to define which creatures are "fittest," except by saying that they are the ones which survive. A similar problem plagues the modern definition of natural selection as "differential reproduction." The problem is that evolutionists have always had an intuitive feeling that the ordinary interactions between living organisms would serve to select those patterns which, over long periods of time, would transform simple molecular arrangements into modern higher life forms. This intuition was based on commonplace examples. For example, a mutation producing longer legs in a deer might well be selected because it would enable the animal to escape from predators more easily. However, this is no reason to believe that na-

31

tural selection would possess the discriminating power needed to guide the development of a world of plants and animals from an inanimate primeval slime. What we have shown here is that in a system governed by simple natural laws, no process, whether it be natural selection or any other imagined principle of evolutionary development, is sufficient to do this.

Let us now examine these arguments further in order to understand the basic principle behind them. First of all, note that the upper bound on M(X) given by inequality (21) will become reasonably large only if we increase the complexity of our system of natural laws to the point where L(M) is nearly equal to 100,000 – 28 – c. Now, 100,000 characters corresponds to about 36 solid pages of coding. Modern science has never reached the point of even considering that nature may run according to laws of this complexity, and it is unlikely that such considerations could ever be practical, given the limitations of the human mind. At any rate, the basic laws of present-day physics are based on the Schrödinger equation and a few basic potentials. They are far less complicated than this.

Of course, L(M) might be increased by increasing the complexity of the initial conditions or the boundary conditions of the system, S. But this would amount to building instructions for the genetic code, X, into the initial or boundary conditions. This would certainly violate the requirements of a theory of evolution. If instructions for X are built into the boundary conditions, then as time passes in S we will at some point find instructions for X coming across the boundary into S. In this case it would be valid to say that X has originated from outside of S and been transmitted into S in some form. But this is not evolution. Likewise, if instructions for X were built into the initial conditions, it would be valid to say that X in some form was already present in S in the beginning. But this is not evolution either. Also, if we code instructions for X into the laws of transformation of S, then we are saying that X is an inherent part of the fundamental absolute nature of things. This also is not evolution.

This point can be made clearer by the following stronger version of the inequality (19):

$$M(X) \leqslant 64^{\,c + \log T - L(X|M)}.$$ (22)

This inequality shows that there is a reasonable probability of finding X in S at time, t, only if L(X|M), the amount of information needed to specify X given M, is not very much greater than c + log T. This constant may be taken to be no more than 60 characters, since c = 32 characters in this case. This essentially means that most of the information needed to code X must already be there in M.

In a study of actual evolution, we would not be dealing with simply one genetic code for one organism. Rather, we would expect to find that many different kinds of plants and animals had evolved, and these may be designated by $X_1, \ldots, X_K$, where K will certainly be in the thousands, if not millions. Since $L(X_k|M)$ would have to be small for each X_k in order for it to have a reasonable probability of being found in S, it follows that $L(X_1 X_2 \ldots X_K|M)$ would have to be fairly small also—certainly no greater than K (c + log T). For a selection of 1,000 organisms, this comes out to 60,000, or less than our minimum estimate of 10^5 for a single higher organism. This means that M would have to contain coding in some form for most of the information required to specify all these organisms.

In this connection the basic inequality,

$$L(X) \leqslant L(X|M) + L(M) \tag{23}$$

indicates that L(M) would have to be quite large, since, as we have pointed out in Section 2, the great diversity of form and function among the different species of life indicates that the total information content, L(X), for a selection, $X = X_1 X_2 \ldots X_K$, of many different species should be very high.

Our essential point, then, is that in a system operating according to a set of physical laws, the only forms at all likely to appear in the course of time are those coded into M. This means that these forms are coded into the initial conditions of S, or that coded instructions for them are transmitted across the boundary at some time, or that they are coded into the laws of transformation themselves. In any case, they cannot be said to evolve; rather, they are built into the system in the first place, or they come from outside.

We would like to emphasize that a complex code picked at random cannot be expected to convey a specific complex message. For example, if cosmic rays enter S across its boundary, and their energies and directions of motion are given by a simple statistical distribution, then the system, S, cannot be expected to extract a large amount of specific information from them in the form of a particular organism with genetic code, X. This is what inequality (22) implies. In order for a large amount of information to be obtainable from the cosmic rays, their statistical distribution function, b, would have to be complex. Also, it could not be just any complex function, but would have to be specifically chosen so that $L(X|b)$ was small. In other words, a large amount of specific information for X would have to be coded into the cosmic ray stream.

This shows that the vague notion that "little bits of negentropy from the sun" can add up over a long time period to yield a living organism of high information content is mistaken. In popular literature advocating the theory of chemical evolution, the negative of the thermodynamic quantity called entropy has been interpreted as a measure of information content. This measure is commonly referred to as "negentropy." Since the biosphere of the earth is thermodynamically an open system, it is possible for the negentropy of a physical entity in the biosphere to increase at the expense of a decrease in the negentropy of the sun, without violating the second law of thermodynamics. This has been interpreted to mean that information is transmitted in the sunlight passing from the sun to the Earth, and that this information can add up to define the structures of living organisms.[22] This argument is fallacious because the negative of entropy is not a measure of information content. For example, a crystal at absolute zero has the highest possible negentropy value (of zero), but it presumably has a very low information content, since a simple program will specify the periodic arrangement of atoms in a crystal lattice. We can see from the analysis of information content that the negentropy argument is wrong. In order for the structural patterns of organisms to be extracted from sunlight, they would in great measure have to be coded into the sunlight in a very specific form.

This possibility is a stronger version of the panspermia hypothesis of Svante Arrhenius. According to this theory, life may never have had an origin, and it may have been spread throughout the universe in the form of spores that could travel in outer space from one favorable planet to another. Arrhenius thought that these spores were primitive and would evolve once they reached their given planet. We can see from this analysis, however, that "primitive" spores would not do—the spores would have to contain instructions for all of the organisms that later evolved. Owing to a variety of serious drawbacks, the theory proposed by Arrhenius never had much standing in science, and these stronger requirements are not likely to make it much more acceptable. Another idea, sometimes discussed in science fiction stories, is that higher intelligences might transmit instructions for organisms across space in the form of coded radio beams.

Both the hypothesis that instructions for the design of living organisms were transmitted across the boundary of S and the hypothesis that they were contained in the initial conditions for S fail to account for the origin of these instructions. Essentially, they amount to the hypothesis that life is eternally existent as a material phenomenon. This suffers from the defect that although the action of simple physical laws cannot, as we have seen, generate information, it can certainly obliterate it. It is common experience that many physical processes tend to destroy order and information, both on a small and on a large scale. This makes it very hard to see how high levels of information content could be maintained eternally in a material system, and there indeed seem to be no theories as yet developed that seriously propose this.

One further consideration involving inequalities (19), (21), and (22) is the following: Are there so many possible varieties within a particular species that even though it is highly unlikely that any one of them would evolve in S, it is nonetheless reasonably probable that at least one out of the total collection would evolve? If that were the case, one could still speak of the evolution of that species as being a possibility. The many varieties of the organism could be more or less trivial variations all possessing the basic features entitling them to be referred to as examples of that species. After all, geneticists have observed that no two individuals of a given species have identical genotypes (with the possible exception of identical twins).

We can easily show, however, that this is not possible. Basically, in order to sufficiently increase the probability of finding at least one genetic code, X, out of our collection, C, of codes, we would have to make C so large that all the members of C could not possibly share in common the basic distinguishing traits of a particular kind of organism. In fact, C would have to be so large that only a very small percentage of its members could even be expected to possess the basic features of life as we know it. Most of the codes in C would simply be arbitrary patterns not corresponding to any kind of viable living organism.

In order to show this, let us first estimate how large C would have to be. Consider a collection, C, of N different varieties of an

35

organism, each of which has a complexity level of $L(X) \geqslant 10^5$ characters. By inequality (21) the probability of finding even one of these varieties in S at time, t, is bounded by,

$$\text{Prob (C)} \ll N\, 64^{-80,000}. \tag{24}$$

In order for Prob(C) to be reasonably large, we would have to have $N > 64^{80,000}$. In comparison, we may note that 10^{75} is a very generous upper bound on the total number of organisms that could have lived on the earth in a 4.5-billion-year-period.[23] Even though life on the earth shows an impressive diversity of form, naturalists have estimated that there are only some two to three million species of living plants and animals (including microorganisms).[24] Also, although one writer has estimated the total number of fossil forms at ten million, only about a hundred thousand of these have been actually described by paleontologists.[25] These figures should provide some perspective on the magnitude of the number, $64^{80,000}$, which we are considering.

Let us specifically examine the subclass, C_K, of codes in C with information content, $L(X|M) = K$. Our bounds, (17) and (18), in conjunction with inequality (23) show that we should consider values of $K \geqslant 100,000 - 16,800$. There are at most 64^K codes in total having the particular information content of K. Yet from inequality (22) we can see that the number, N_K, of codes in C_K must be

$$N_K \geqslant 64^{\,K-60-A}, \tag{25}$$

in order to have Prob(C_K) as great as 64^{-A}. What this means is that in order for Prob(C_K) to be reasonably large, the number of codes in C_K must be of nearly the same order of magnitude as the total number of codes having information content, K.

Should we expect such a large proportion of the codes of information content, K, to specify living organisms? We can easily see the answer to be no. Consider that for each viable organism there are vast numbers of defective forms that are not viable. Each gene code for a viable organism is subject to many thousands of different fatal mutations; and n kinds of fatal mutations can be combined in 2^n different ways, most of which should specify nonviable forms. Beyond this there should be enormous numbers of forms which do not correspond to any living organism at all, and most of which follow no coherent pattern that would convey any meaning to us or appear at all familiar.

The class, C_K, should therefore have a substantially lower order of magnitude than the total class of codes of information content, K. Since C is composed at most of a few billion C_K's for K's between our lower and upper bounds on genetic information content, it follows that Prob(C) must be very small if C is restricted to the class of life forms. [Prob(C) is the sum of the different Prob (C_K)'s.]

Further light is shed on this matter by the concept of information shared "in common" by a set of codes. We would expect all the members of a particular category of organisms to share in common certain information which defines that category of form. For example, if there is any meaning to the term "horse," there must be certain specific information common to all horses and not shared fully with any other forms. The arguments in Section 2 indicate that for a higher animal, this characteristic defining information should have a content of at least 10^5 characters.

For a class of forms of fixed information content, there must be a trade-off between the number of forms in the class and the amount of information they can share in common. The totality of codes of information content, K, share no information in common, and subclasses of nearly the same order of magnitude as this class (such as C_K) can share very little. For this reason, we would not expect such subclasses to correspond to well-defined complex categories of forms, such as "horses" or even "vertebrates."

The concept of shared information is formally defined in the appendix. The following inequality gives a bound on the probability of finding a configuration, X, belonging to a certain class, C, in the physical system at time, t.

$$\text{Prob(Some X in C exists in S at time, t)} \leqslant 64^{c' + \log T + L(M) - K}. \quad (26)$$

Here it is assumed that C is a collection of codes, all of which possess K characters of specific information in common. The class of horses should share in common the information for all the essential features that go into the structure of a horse. As described in Section 2, this includes information for eyes, legs, circulatory systems, etc., and should have an information content of at least 10^5 characters. Substituting 10^5 for K, we can conclude that no representative of this class of forms is likely to evolve.

Thus, the chances that a particular form or reasonably close variant of that form will evolve within a physical system are extremely small unless the information specifying that form is essentially built into the initial conditions, boundary conditions, or natural laws of the system.[26] Multiplying by 10^{75} (our upper bound on numbers of organisms), we can conclude that the chances are exceedingly small that any life form will evolve which has ever existed on the earth in 4.5 billion years and which has a greater information content than $L(M) + \log T + 75\log 10 + c'$.

4

What is the origin of life?
Some implications of this analysis.

We have argued that the information for the structures of living organisms must either be built into the initial and boundary conditions of the system, or must be built into the system's fundamental laws. In the first case, we are left with the question of where the information came from. It must have had some origin, either on the other side of the boundary or at a time previous to the initial time, t_0. In the second case, we are confronted with a system of absolute, invariant laws that encode all of the detailed information for thousands of species of living beings, for human civilizations, symphonies, etc. In either case, we have a world picture with a very dissatisfying degree of incompleteness or disunity. Either the information for the intricate and harmonious world we perceive is coming from an unknown source, or it is coded into a kind of abstract cosmic laundry list of laws that satisfy (due to their very irreducibility) no unifying principle.

We have already rejected the alternative that the information for life forms originated in the boundary or initial conditions of the system. Essentially, this alternative leaves us with an infinite regress that is worse than no explanation at all: as we push the boundaries of the system further and further back in space and time we must continue to find the same intricate information coded in some kind of material configurations. This picture of things is especially unsatisfactory because the transformations of matter occurring in nature have a strong tendency to obliterate information. We will not consider this picture further here.

The other alternative is that the information is built into the fundamental natural laws. It is more or less typical for scientists to believe that the laws they know at the moment are the ultimate laws—and this is especially prevalent today in the life-sciences. There is, however, much evidence that the laws known to science today are by no means complete or final. Even in the realm of biochemistry there is increasing evidence of phenomena that do not fit into the current theories. These phenomena give support to the idea that many laws operating in nature are still unknown.

For example, the research of C. L. Kervran in France has

39

shown that many chemical elements can be created and destroyed within the bodies of living organisms in the course of their ordinary metabolism.[27] He describes these phenomena as "biological transmutations" wherein organisms can transform one element into another by nuclear reactions without resorting to the high energies and temperatures these reactions require according to our present understanding. The transmutations are of interest because they directly indicate the existence of as yet unknown laws which are intimately involved with the phenomena of life. This analysis implies that the search for such laws should prove to be a very fertile field of future scientific investigation. Far from their being "impossible," we should expect to find many such remarkable phenomena. (That biochemists have failed to observe these phenomena in spite of their extensive chemical analyses of living organisms, demonstrates the blinding effect that narrow adherence to one limited hypothesis can have on human understanding.)

Let us explore more thoroughly the idea that the information for life forms is stored in an extensive system of what could be called "higher order laws." Fundamental laws are conceived of in modern science as being invariant in time and all-pervading in space. This is somewhat mysterious, but it becomes more so if we require that these laws encompass large amounts of mathematically irreducible information specifying living organisms, cultures, and so forth. We must suppose that large numbers of invariant, all-pervading laws in some sense "exist." Yet the very fact that they possess a high information content means that they cannot be mathematically reduced to any unified scheme.

We are thus confronted with an invisible and intractable collection of absolute rules that just so happen to specify a variegated world of living beings. This alternative is also unsatisfactory, yet we cannot avoid it as long as we adhere to the basic mathematical approach to the study of nature outlined in Section 1.

We are now forced to reject both of our original alternatives for explaining life as entirely unsatisfactory. Neither constitutes what we could call an explanation. Is there any other recourse? We suggest that a reasonable and satisfying alternative can be found only if we are willing to go beyond the strictly mathematical approach to nature. We can do this by introducing "life" as a basic principle, in and of itself, which is not capable of exhaustive mathematical description.

The idea is that "life" is something irreducible—something that cannot be explained in terms of combinations of other non-living things. An analogy would be the concept of electric charge, which in current physical theories is also a fundamental entity that cannot be expressed in terms of anything else. In this picture, the laws of nature cannot be finally represented by fixed mathematical equations. This doesn't mean, of course, that equations cannot exist which approximate these laws. We may expect to find many different laws operating in nature. However, these should exist as derived entities depending ultimately on the higher, nonmathematical principle of life. This picture also requires that the physical combinations of elements which form the bodies of living organisms—and which are commonly regarded as constituting life itself—must be regarded instead as byproducts of life.

Since our analysis shows that the final cause underlying the phenomena of nature must contain information for all the forms of living beings, we propose that this final cause is itself a primordial living being lying completely outside the realm of mathematical describability. All of the temporary, numerically describable patterns of the material manifestation arise from this being. The individual living entities found in nature may be considered minute quantized parts of this being, just as electrons are quantized particles of electric charge. As certain electrical properties are characteristic of electric charge, likewise complex and yet harmonious activity—that is, intelligence—is characteristic both of this primordial being and of the derived quantized parts.

We are considering here the question of what is the character of the absolute truth, or the fundamental, underlying cause of phenomena. The driving motive behind science—at least in its purer forms—has always been to arrive at an understanding of this ultimate cause. (This is of course denied by the philosophy of positivism, which holds that science is and can be nothing more than the search for regular patterns in the welter of sense data.[28]) The implication of our analysis is that no coherent, unified view of nature is in fact possible as long as we accept the hypothesis that nature is mathematically describable. Although this hypothesis has dominated science for the last two to three hundred years, we would like to suggest that it is, in fact, unnecessary. We are proposing that a valid and fruitful alternative exists which circumvents the frustrating impasse inherent in this viewpoint.

From the point of view of scientific procedure, it is perfectly valid to consider our hypothesis, especially if we hope to obtain actual knowledge of reality from scientific study. Even from the point of view of positivism, this hypothesis may be regarded as an admissible proposal that will bear implications for the expected ordering of sense data. With a strictly numerical approach to empirical data, we can only hope to deal with numerical patterns and mathematical relationships. From such a viewpoint, the primordial living being can be manifest only in terms of more and more complicated and irreducible patterns of material interaction. Here, then, is one falsifiable prediction of this hypothesis: we may expect to find unlimited patterns and relationships in the world that defy reduction to any finite scheme. (This, of course, has been the gist of this paper.)

Philosophically, the problem of unity in nature is solved by the proposal that the absolute truth is a purposeful living being. Traditionally, scientists have felt that there is unity and harmony in nature and have sought to find this unity in the extreme simplicity of the natural laws. Thus, Einstein sought to find a unified field theory, which would express the laws of both gravity and electromagnetism in one formula. Such unity is ruled out if the natural laws have a high level of complexity.

However, this sought-after unity can be attained if we consider sentient life itself irreducible and absolute. The harmony in diversity lies in the purposeful comprehension and direction of the variegated cosmic manifestations by the ultimate sentient being. We are dealing here with the question of what should be regarded as the irreducible absolute truth: we are basically forced to choose between a senseless multitude of arbitrary rules and conditions and a higher unifying principle lying (necessarily) outside the grasp of mathematical expression. That this higher principle should have the character of life follows from simple empirical observation: we have been led to the consideration of complex laws and conditions from the observation of life.

One of the basic aspects of the reductionistic approach to understanding nature is that it eliminates "meaning" and "purpose" from the picture altogether. According to this approach, simple physical causation is to be put in the place of intelligent purpose. However, since we have eliminated *simple* physical causation as a possible cause of life, we propose that "meaning" and "purpose" should be reinstated. This requires the existence of conscious intelligence as the fundamental causal principle in nature. This not only makes sense out of the otherwise meaningless display of complex form without purpose, but it also ties in the theme of consciousness in a unified way.

The phenomenon of consciousness fits neatly into place if we suppose that life is an absolute principle. In Western scientific thought there has been great difficulty in fitting consciousness into nature in any unified way. The standard view of science has been that consciousness is simply a phenomenon of material inter-action—or a dangling epiphenomenon. However, no one has ever been able to indicate how the interaction of insentient entities can produce conscious awareness. In fact, consciousness seems to have an inherently irreducible nature.

This problem is solved if we accept conscious self-awareness as an inherent feature of the irreducible absolute truth. Conscious-ness can then be seen to occupy a natural role as a fundamental unifying characteristic of the absolute living being. The individual-ized consciousness of the multitudes of living beings can be under-stood in a simple and unified way if these beings are regarded as minute quantized parts of the single absolute being. These minute parts derive all of their properties, including that of conscious awareness, from this absolute source. The bodies of the living organisms may be regarded as carriers of these "quanta of life."

In this paper our basic argument, drawn from information theory, supports the understanding that the unified absolute cause lying behind nature must be an irreducible living being. The other concept we have introduced—that of a class of quantized individ-ual parts of this being—does not directly follow from this argu-ment. One might suppose that the one, absolute living being had simply generated a series of automatons out of material elements. However, the concept of the quantum of life follows naturally from these considerations: Many individual conscious beings (such as ourselves) *do* exist, and consciousness is not reducible to a combination of material elements. Since consciousness plays a natural unifying role as an inherent feature of the absolute truth, the many conscious beings may be most simply seen as its quan-tized parts.

A more detailed analysis of consciousness will be presented in other papers of this series. We would simply like to stress here that all of these concepts—far from being unscientific, or even anti-scientific—are potential subjects for truly scientific investigation and understanding. We might also point out that we are not vio-lating the principle—known as Occam's razor—that one should not multiply entities beyond necessity. We feel that the picture we have outlined here is more coherent and economical than the alter-native picutre of thousands of piecemeal laws.

As a final point, we should say something about the nature of chance. Thus far we have referred to "chance" without giving it any explicit definition, and we have manipulated it mathematically according to standard formulas. In this way we have calculated that the chance that a given form will evolve goes down exponentially with the difference between its information content and the information content of the system.

What is the meaning of chance? There are two basic conceptions of chance. According to the first, chance is a measure of the ignorance of an observer about some event. According to this view, the event itself is perfectly definite, but the observer refers to it in terms of chance because his knowledge is imperfect. In this view, chance, is subjective and may vary from one observer to another. According to the second view, there actually exist chance events, which are by nature undetermined and which obey only statistical laws.

The first viewpoint is represented by the so-called classical theory of probability, in which equal chances are assigned to alternative events when there is no reason to expect one of them to happen in favor of another. This is clearly a means of describing one's ignorance numerically. The second viewpoint is represented by the relative-frequency interpretation of probability, in which the probability of an event is measured by repeating the circumstances of the event many times and seeing how often it comes up. The standard example is the flipping of a coin. Here it is often supposed that the event itself is random. An example is the interpretation of the state functions of quantum mechanics, which are believed to give an inherently random description of phenomena such as radioactive decay and the movement of atomic particles.

The analysis presented here strongly favors the first point of view, since the inequalities show a reciprocal balance between information content and probability. As the gap between the information needed for Y and the information provided by the system goes up, the probability of finding Y correspondingly goes down. Thus, probability can be seen as measuring a lack of information.

It is therefore interesting to note that in an analysis of several standard theories of probability, the mathematician T. L. Fine found that none gave a reasonable characterization of the second interpretation of probability.[29] Circular reasoning and unavoidable contradictions were found in each proposed method of describing "chance" as an actual phenomenon of nature. Thus, only the first point of view seems to provide a tenable interpretation of chance.

According to the world picture presented here, we expect all phenomena to be of an essentially deterministic character. The statistical nature of the present quantum mechanical laws can be seen as a symptom of their incompleteness. In general, we expect that natural laws can be only approximately described by mathematical equations, and it is not surprising for the mathematical approximations to contain a statistical element reflecting their incomplete nature. The laws themselves can be understood as manifestations of the primordial living being.

Thus we conclude that life exists as the absolute source of the material manifestation, and that the process by which the physical forms of living entities are created is quite different from the process envisioned by the theory of evolution. The slogan of evolutionary theory has been, "Life comes from matter." This analysis shows that this is wrong. The real slogan should rather be: "Both life and matter come from life."

Appendix.
Mathematical derivations.

In order to define information content, we must first establish a fixed "computer," C. By

$$C(P|f_1, \ldots, f_n) = X \tag{27}$$

we mean the following: The computer, C, executes the program, P, yielding an output of X. The functions, $f_1, \ldots, f_n$, are externally supplied functions that may be used by the program, P, in the course of its calculations (subroutines). X may be either a number or a function. If X is a function, it is understood that the arguments of X must be supplied as inputs to the program, P, which calculates X.

We shall assume that all programs are to be written in a standard programming language with an alphabet of 64 characters. We will specify a number of special features of this language as we need them. Programs can be coded as binary numbers. If P is such a number, *define* $1(P)$ to be the number of binary digits in P (including leading zeros.) We will refer to the lengths of programs both in characters and in binary digits, or "bits." (One character equals 6 bits.)

For the purpose of calculating solutions to equations arising in physics, the programming language may be assumed to have facility for expressing standard mathematical operations. It is also convenient to assume that this language can refer to numbers with arbitrarily large numbers of digits.

Define the *information content*, $L(X|f_1, \ldots, f_n)$, of X given $f_1, \ldots, f_n$ to be the length of the shortest program (or programs) that will compute X, given that $f_1, \ldots, f_n$ are available to refer to during the calculation.

In the previous sections, we measured information content, L, in characters, since this seems more easy to visualize. In this section we shall always measure L in binary bits, unless otherwise mentioned. This is more convenient in the mathematical derivations.

This method of measuring information content was first devised by the Russian mathematician Kolmogorof.[30] The subject has been developed further by G. Chaitin.[31] Another name we shall use for information content is *complexity*.

The first thing to note about L is that relatively few X's have a low information content. The number of X's with information content, $L(X|f_1, \ldots, f_n)$, less than k is no more than 2^k. This is because there are no more than 2^k programs, P, with $1(P)$ less than k. It is interesting to note that this remains true no matter how many functions, $f_1, \ldots, f_n$, are available for reference by P.

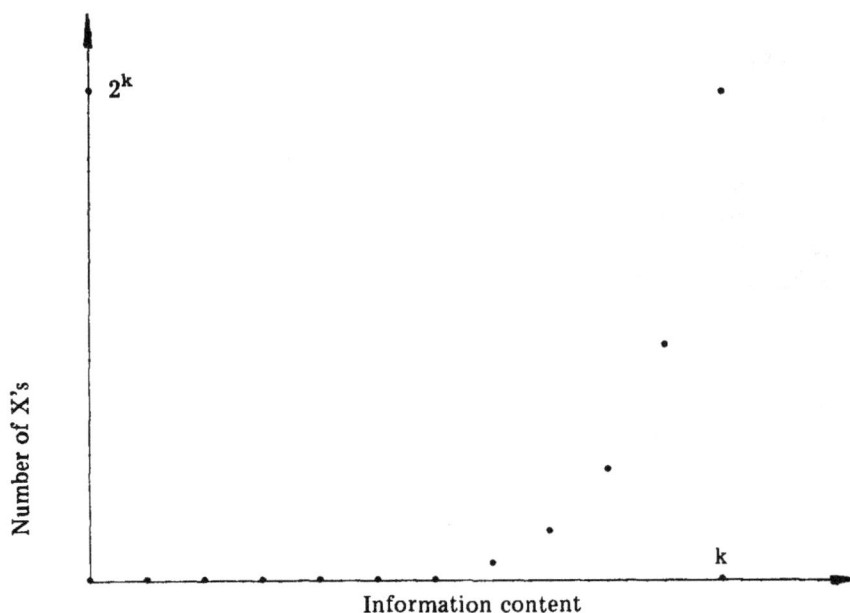

Figure 5. Upper bound on the number of X's with information content less than k.

Define $s[R(\cdot),w]$ to be the wth number, X, in numerical order, for which the statement, R(X), is true. This is considered to be defined only if there is such an X. We shall assume that the function, $s(\cdot,\cdot)$, which is defined in this way may be referred to in the programming language. (It could otherwise be expressed by means of a "do loop.") R(X) must be a statement that can be expressed in the computer language. We shall be interested in statements of the form, $A \geq B$, where A and B are numbers or expressions for the calculation of numbers.

47

We shall need some provision in the programming language for coding information in the most compact way possible. One way of doing this is to adopt the following convention for the coding of positive integers: A positive integer, j, can be coded in the form, $n; b_1 b_2 \ldots b_n$, where n is the number of binary bits (0 or 1) needed to write j in binary notation and $b_1 b_2 \ldots b_n$ is the actual binary expression for j. Both n and the semicolon are coded in terms of characters, each of which requires six binary bits to code. This enables us to encode j in such a way that if $j < 2^m$ then

$$L(j) \leqslant m + L(m) + 6. \tag{28}$$

The '6' corresponds to the coding for the semicolon.

We shall also need a method whereby programs can be joined together. Let us make the convention that a program must be in the form of a numerical constant, xx...x, a function, f(...), or a string of symbols surrounded by parentheses, (...). We shall call a function or variable symbol in a program *free* if it is used but not calculated in the program. Suppose x is a free function or variable symbol appearing in the program, P. P might be of the form, (...x...x...x...), with several appearances of x. If x is a variable symbol, let Q be a program that calculates a numerical constant, and if x is a function symbol requiring n variables, let Q be a program computing a function of n variables. Interpret (...Qx...x...x...) as the program obtained by executing P and using Q to compute a value for x whenever x is used. This notation unambiguously establishes Q as a subroutine for P.

As a result of this convention we have the basic relationship,

$$L(Y) \leqslant L(Y|X) + L(X). \tag{29}$$

Using these conventions, we can derive the following basic theorem:

Proposition 1. Let M(X) be a function mapping non-negative integers into non-negative real numbers for which,

$$\sum_X M(X) \leqslant T. \tag{30}$$

48

Then it follows that,

$$M(X) \leqslant 2^{c \,+\, \log_2 T \,-\, L(X|M)}.$$ (31)

The constant, c, is no more than 191 bits, or 32 characters.

Proof Let k be a positive integer, and let N be the number of X's for which $M(X) > 2^{-k}T$. By equation (30) we know that $N(2^{-k}T)$ is less than T. It follows that if $M(X) > 2^{-k}T$, then we can write,

$$X = s[M(\cdot) > 2^{-k}T, w]$$ (32)

for some $1 \leqslant w \leqslant N < 2^k$. If we write $k; b_1 b_2 \ldots b_k$ for w, this gives us an expression for X which requires 17 characters (including multiplication and exponentiation symbols) and k binary bits and which refers to k, T, and M as unknowns. Therefore,

$$L(X|M,k,T) \leqslant 6 \times 17 + k.$$ (33)

This further implies that,

$$L(X|M) \leqslant 6 \times 17 + L(T) + L(k) + k.$$ (34)

Let $b = L(X|M) - L(T) - 103$. Write $L_m(j) = \max [L(1), L(2), \ldots, L(j)]$. If we pick $k = b - L_m(b)$, then it follows that $L(k) + k \leqslant b$. Since $103 = 6 \times 17 + 1$, this means that (34) is violated for this choice of k. This means that $M(X) \leqslant 2^{-k}T$, or

$$M(X) \leqslant 2^{\,103 \,+\, L(T) \,+\, L_m(b) \,+\, \log_2 T \,-\, L(X|M)}.$$ (35)

For the constant, c, let us pick some number bigger than the sum of the first three terms in this exponent of 2. We expect to take a value of 10^{51} for T, in which case $L(T) = L(10^{51}) = 5 \times 6$. We will not be worried about X's with L(X) much bigger than about 2^{40}. If $b < 2^{40}$, then $L_m(b) \leqslant 3 \times 6 + 40$. Adding these together, we get c = 191 bits, or less than 32 characters. Q.E.D.

Corollary 1. Suppose that the function, M, is the same as in proposition 1. Then,

$$M(X) \leqslant 2^{c + \log_2 T + L(M) - L(X)}.$$ (36)

Proof This follows from proposition 1 and (29). Q.E.D.

Corollary 2. Suppose that M is the same as in proposition 1. Suppose that G is a function mapping non-negative integers to non-negative integers. Then,

$$\sum_{X:\ G(X)\ =\ Y} M(X) \leqslant 2^{c' + \log_2 T + L(M) - L(Y|G)}.$$ (37)

Proof Define $M_1(Y)$ to be the left-hand side of (37). Then,

$$\sum_Y M_1(Y) = \sum_X M(X) \leqslant T.$$ (38)

If we perform the derivation of (36) with G as an externally supplied function, we obtain:

$$M_1(Y) \leqslant 2^{c + \log_2 T + L(M_1|G) - L(Y|G)}.$$ (39)

The left-hand side of (37) gives us an expression for M_1 in terms of M, G, and c_1 bits worth of additional programming to express the sum. This means that $L(M_1|G) \leqslant c_1 + L(M)$. The desired inequality follows if we take $c' = c_1 + c$. Q.E.D.

Definition. The numbers, $X_1, \ldots, X_m$, shall be said to possess K bits of information *in common* if there is a function, G, such that $G(X_1) = G(X_2) = \ldots = G(X_m) = Y$, and $L(Y|G) \geqslant K$.

The idea behind this definition is that the function, G, represents a program or process of analysis intended to extract information from the X's. If the same Y can be extracted from all of the X's, then they are said to possess in common the information in this Y. $L(Y|G)$ is used to measure the amount of this information instead of $L(Y)$ because no information that Y obtains from G itself could be said to belong to the X's.

In the reductionistic view of the world, according to which a human being is regarded as a computing machine, the action of a human observer may be taken as an example of the function, G.

Thus, a series of books would be said to have the statement Y in common if an observer would be able to write down Y after examining any one of the books. Of course, any aspects of Y originating in the observer himself might not be actually shared by these books.

Using this definition, Corollary 2 can be rewritten as follows: If $X_1, \ldots, X_m$ have K bits of information in common, then,

$$M(X_1) + \ldots + M(X_m) \leqslant 2^{c' + \log_2 T + L(M) - K}. \qquad (40)$$

Here is yet another way of looking at this basic situation: Suppose that R(X,Y) is an equivalence relation on the nonnegative integers. For example, if X and Y are regarded as descriptions of living organisms, then R(X,Y) might mean "X and Y both belong to the same species of life." The relation, R, divides the X's into disjoint equivalence classes, $C_1, C_2, \ldots, C_n$. Let us define the *simplest member* of the class, C_j, to be that X in C_j which has the smallest value of L(X|R). (If there are several such X's, pick the numerically smallest.) Also define the *minimum amount of information* in C_j to be equal to this value of L(X|R). Call this I_j. In our example, I_j is the information content of the simplest individual organism in the species, C_j. We use L(X|R) for this instead of L(X) because this information should be independent of the process of observation, R, which determines whether or not two individuals belong to the same class.

The following theorem holds for these equivalence classes:

Corollary 3. Let M be the same as in proposition 1, and let R be an equivalence relation, as described above. Then for each j,

$$\sum_{X \text{ in } C_j} M(X) \leqslant 2^{c'' + \log_2 T + L(M) - I_j} \qquad (41)$$

Proof Define G(X) = s[R(·,X),1]. Then R(X,Y) holds if and only if G(X) = G(Y). It follows that $I_j \leqslant L(Y|R) \leqslant L(Y|G) + c_2$ for Y in C_j. By corollary 2 and $c'' = c' + c_2$ we obtain the inequality, (41). The constant, c_2, is the number of bits needed to express R in terms of G. This can be done by the statement, "R(X,Y) iff G(X) = G(Y)." Q.E.D.

Let us briefly indicate some of the applications of proposition 1 and its corollaries. Consider a quantum mechanical system with variables for the position of N atomic nuclei corresponding to a variety of different elements. (There will also be spin variables, electromagnetic variables, and variables for the positions of electrons.) The bonding together of various atoms to form molecules can be expressed in terms of the relative distance between these various nuclei. A necessary condition that two atoms be bonded together is that their corresponding nuclei be situated within a certain distance of one another. This distance will depend on the type of atoms and bond being considered. The pattern of bonds making up a particular molecule can be encoded into the digits of a binary number, X, in some standard way. Using this coding scheme and the data for interatomic bonding distances, one can define a function, $B_X(\bar{q}_1, \ldots, \bar{q}_N)$, which will equal 1 if the nuclear positions, $\bar{q}_1, \ldots, \bar{q}_N$, are so arranged that some $\bar{q}_{i_1}, \ldots, \bar{q}_{i_k}$ among them satisfy the spacing conditions necessary for the molecule described by X. This function will otherwise be set equal to 0.

The function, B_X, provides us with an operator that enables us to calculate the probability, $M(X, \psi)$, that the molecule described by X exists within the physical system when that system is in the quantum mechanical state, ψ. (Or, rather, it provides us with an upper bound on this probability, since we are dealing only with a necessary condition for the existence of the molecule.) Specifically, we have,

$$M(X, \psi) = \int \psi^*(\bar{Q}) \, \psi(\bar{Q}) \, B_X(\bar{q}_1, \ldots, \bar{q}_N) \, d\bar{Q}, \qquad (42)$$

where the integral is taken over all of the variables of ψ.

For any given arrangement of the variables, $\bar{q}_1, \ldots, \bar{q}_N$, there will be only a limited number of molecular configurations, X, that can be found among them. That is, there should be a number, T, for which,

$$\sum_X B_X(\bar{q}_1, \ldots, \bar{q}_N) \leqslant T \qquad (43)$$

for all arrangements of $\bar{q}_1, \ldots, \bar{q}_N$.

In order to obtain an estimate of T, let us make one further restriction of B_X. Suppose that $B_X(\bar{q}_1, \ldots, \bar{q}_N)$ shall equal 1 only if there is a collection, $\bar{q}_{i_1}, \ldots, \bar{q}_{i_k}$, of these variables that satisfy the conditions for X but are not within bonding distance of any other $\bar{q}_i$'s. Then $B_X = 1$ will mean that the molecule represented by X is present without being part of a larger molecule. This means that each $\bar{q}_i$ can be part of at most one configuration corresponding to an X with $B_X(\bar{q}_1, \ldots, \bar{q}_N) = 1$. Therefore, $T \leqslant N$. In order to estimate N, let us consider that Avogadro's number, the number of molecules in one molecular weight of a substance, is about 6×10^{26}. For the entire earth we can estimate that there will be no more than 10^{51} atoms. These values indicate what we might expect of T in biological applications of proposition 1.

The function, $M(X,\psi)$, can be calculated using (42) and the calculations for ψ in terms of natural laws, initial conditions, and boundary conditions. We can also calculate the probability of finding X in a system described by a mixture of states. If the mixture consists of the states, ψ_i, with probability, a_i, then this probability is given by $\Sigma\ a_i M(X,\psi_i)$. Thus, the total information content of the function, $M(X)$, which bounds the probability of finding the configuration, X, in the system at a given time, t, is given by:

$$L(M) \leqslant L(\text{initial conditions}) + L(\text{boundary conditions}) + L(\text{natural laws}) + L(t) + L(B) + \text{constant}. \tag{44}$$

The constant equals the number of symbols needed to express (42), plus some other odds and ends. B refers to B_X, with X considered as a variable. The time, t, is counted since it enters into the calculation of the state of the system. $L(t)$ will be no more than the number of characters needed to express t in base 32. This is small for t ranging from 0 to billions of billions of years.

Our next step is to derive a theorem that will make it possible to estimate a lower bound for $L(X)$. *Define* the function, $I(n,X)$, as follows: $I(n,X)$ is the number of programs, P, for which $C(P) = XY$ for some Y, and $1(P) = n$. By XY we mean the concatenation of the two strings, X and Y. For completeness we also define $I(n,\emptyset)$ to be the number of programs, P, for which $C(P) = Y$ for some Y, and $1(P) = n$. Clearly, $I(n,\emptyset) \leqslant 2^n$.

Proposition 2. If X and Y are two non-negative integers, then

$$I(n,XY) \leqslant I(n,X)\ 2^{c\ -\ L(Y|X,I,n)}. \tag{45}$$

The constant, c, is no more than 43 characters.
    Proof It follows from the definition of I that,

$$\sum_{Y:\ 1(Y)\ =\ h} I(n,XY) \leqslant I(n,X). \tag{46}$$

Define $M(Y) = x(1(Y),h) I(n,XY)$ and $T = I(n,X)$. Here $x(a,b)$, the indicator function, is taken to be 1 if $a=b$ and 0 otherwise. We shall assume that the computer language can refer to this function.

With these definitions we can see that the conditions of proposition 1 apply. If we carry out the steps of the proof of proposition 1, we find that inequality (35) becomes

$$I(n,XY) \leqslant I(n,X)2^{L_m(b) + 6 \times 39 + 1 - L(Y|X,I,n)} \tag{47}$$

as long as $1(Y) = h$. Here we assume that symbols for concatenation of strings and exponentiation to powers of 2 are included in the programming language. We also assume that the language can handle numbers of base 32, and that h is expressed by a number of this kind of no more than four digits. As we are interested in cases where $L(Y|X,I,n) < 32^4$, we may take four characters for $L_m(b)$. This gives us a figure of about 43 characters for c.[32] Q.E.D.

Now, suppose that we have $Y_1, \ldots, Y_N$, with $1(Y_1), \ldots, 1(Y_N) < 32^4$. Let $X_j = Y_1 Y_2 \ldots Y_j$, for $j = 1, \ldots, N$. Write $X_0 = \emptyset$.

Corollary 4. Let $n = L(X_N)$. Then,

$$L(X_N) \geqslant \sum_{j=1}^{N} [L(Y_j|X_{j-1},I,n) - c]. \tag{48}$$

Proof By proposition 2,

$$I(n,X_N) \leqslant I(n,\emptyset)\, 2^{\sum_{j=0}^{N-1} [c - L(Y_{j+1}|X_j,I,n)]} \tag{49}$$

as long as $L(Y_{j+1}|X_j,I,n) \leqslant 32^4$. Since $n = L(X_N)$, we have $I(n,X_N) \geqslant 1$. We also have $I(n,\emptyset) \leqslant 2^n$. As a result, n plus the exponent of 2 in (49) must be non-negative. Q.E.D.

The purpose of corollary 4 is to provide a means of estimating a lower bound for $L(X_N)$ for a long string of digits, X_N. The idea is to break X_N into a number of segments, Y_j. Corollary 4 indicates that if $L(X_N)$ is very low, then the average value of $L(Y_j|X_{j-1},I,n)$ must be very close to c. This means that Y_j can be calculated from X_{j-1} by means of a program using n and I which can be written down in little more than $c = 43$ characters. If $n = L(X_N) < 32^4$ then we can write,

$$L(X_N) \geqslant \sum_{j=1}^{N} [L(Y_j|X_{j-1},I) - c'] , \tag{50}$$

where $c' = c + L(n)$ is no more than 4 characters larger than c. This is the equation we used in Section 2.

If X_N is the genetic coding sequence for a living organism, especially a higher plant or animal, then the Y_j's will correspond to genetic coding for different organs, tissues, and other features of the organism. Since it seems highly implausible that these features can be mapped into one another by transformations requiring no more than about c symbols to write down, we conclude (as described in Section 2) that $L(X_N)$ must be quite large. This certainly agrees with the intuition that the living organisms must be highly complex in structure.

Let us examine this arrangement in greater detail. *Define*

$$K(X,Y) = \min \{ k: I(n,XY) > 2^{-k}I(n,X) \} . \tag{51}$$

It follows that,

$$I(n,X_{j+1}) \leqslant I(n,X_j) \, 2^{1 - K(X_j,Y_{j+1})} . \tag{52}$$

If we let $K_j = K(X_{j-1},Y_j)$, then the reasoning used in the proof of corollary 4 gives us:

$$L(X_N) \geqslant \sum_{j=1}^{N} (K_j - 1) , \tag{53}$$

when we choose $n = L(X_N)$.

Define the function, $F_w(h,n,X)$ by,

$$F_w(h,n,X) = s[x(1(\cdot),h) \times I(n,X\cdot) > 2^{-1(w)} I(n,X),w] . \tag{54}$$

The dots $(\cdot)$ indicate that $Y = F_w(h,n,X)$ is the wth Y in order satisfying the indicated inequality. As we have pointed out in the proofs of propositions 1 and 2, there are less than 2^k Y's for which $1(Y) = h$ and $I(n,XY) > 2^{-k}I(n,X)$. If k is chosen to be $k = K(X,Y)$, then this inequality is satisfied. Therefore, if we are given X and Y with $1(Y) = h$, we can write,

$$Y = F_w(h,n,X) , \tag{55}$$

55

where w is a binary integer of k = K(X,Y) digits. [This may include leading zeros. We want $1(w) = k$ in (54).]

Summing this up, we obtain the following result:

Proposition 3. Suppose that $1(Y_1) = 1(Y_2) = \ldots = 1(Y_N) = h$ and let $X_j = Y_1 \ Y_2 \ldots Y_j$. Then there are numbers, $K_1, \ldots, K_N$ and binary integers, $w_1, \ldots, w_N$ for which $1(w_j) = K_j$ and,

$$Y_j = F_{w_j}(h,n,X_{j-1}), \tag{56}$$

and

$$L(X_N) = n \geqslant \sum_{j=1}^{N} (K_j - 1). \tag{57}$$

Proof This is proven above. Q.E.D.

From (51) we can see that K(X,Y) must be positive, since $I(n,XY) \leqslant I(n,X)$. Therefore each K_j in proposition 3 must be positive. Let us consider what happens when $L(X_N) = aN$ for some $0 < a \leqslant 1$. This means that the average value of K_j will be no more than $a+1$. This implies that the number of K_j's which equal 1 must be at least $N(1-a)$. For these K_j's we will have $w_j = 1$, since w_j cannot be zero. It follows from this that at least $N(1-a)$ Y_j's will be given by $F_1(h,n,X_{j-1})$. If some of the K_j's are very large, even larger numbers of them must equal 1 in order to preserve the average, and hence even more Y_j's will be generated by this formula. Also, if $K_j = 2$, then Y_j is given by F_1, F_2, or F_3 in (56).

Let us apply these results to the case where X_N is the genetic coding for an organism and the Y_j's are a series of segments of this code. It certainly seems unlikely that so many Y_j's could be generated from X_{j-1} by such a simple transformation as F_1. F is given by equation (54), and I is defined just before proposition 2. The function, I, depends only on the properties of the computer language, and so it seems quite unreasonable to suppose that it could have the property, in conjunction with equation (54), of transforming the genetic coding for, say, the liver into the coding for the eye. For this reason, we conclude that $L(X_N)$ cannot be very small.[33]

Notes

1. Watson, *Molecular Biology of the Gene*, pg. 105.
2. Watson, pg. 67.
3. *Newton's Principia*, pg. lxviii.
4. von Helmholtz, *Über die Erhaltung der Kraft*, pg. 6.
5. The quantum mechanical equations are taken from von Neumann, *Mathematical Foundations of Quantum Mechanics*, and Messiah, *Quantum Mechanics*.
6. Fisher, *The Genetical Theory of Natural Selection*.
7. See, for example, Oparin, *The Origin of Life* and Orgel, *The Origins of Life*.
8. This is discussed in Tolman, *Statistical Mechanics*.
9. von Neumann, *Theory of Self-Reproducing Automata*.
10. The data for *E. Coli* is taken from Watson, chap. 3.
11. Watson, pg. 85.
12. Watson, pg. 507.
13. Watson, pg. 508.
14. Watson, pg. 461.
15. Watson, pg. 540.
16. Watson, pg. 94.
17. Satir, "How Cilia Move," *Sci Amer*, Oct. 1974.
18. Berg, "How Bacteria Swim," *Sci Amer*, Aug. 1975.
19. Macbeth, *Darwin Retried*, pg. 101.
20. Alfred Russel Wallace, *The World of Life*.
21. Specifically, $G_w(X) = F_w[4200, L(X_N), X]$, where F is given in equation (54) of the appendix. [$L(X_N)$ should be in bits here.]
22. This argument can be found in Orgel, *The Origins of Life*.
23. This is more than the number of cubic Ångström $\times$ microseconds in the volume of the biosphere $\times$ 4.5 billion years.
24. Percy E. Raymond, "The First Animals and Plants" in Preston Cloud, ed., *Adventures in Earth History*, pg. 667.
25. Normal D. Newell, "The Nature of the Fossil Record" in Preston Cloud, ed., *Adventures in Earth History*, pg. 649.
26. This neglects the small constants, c, c', etc., which appear in the various inequalities. These are small and leave the basic conclusion unaffected. They are artifacts of the mathematical derivations and do not seem conceptually necessary.
27. Kervran, *Biological Transmutations*.
28. The doctrine of positivism holds that one can speak meaningfully only of sense perceptions, and that all other categories are mere verbiage. This view is described by the physicist Y. Freundlich as follows: "To us, the statement that trains have wheels when they are not in the station (when we are not sensing them) *means* that at the station they will have wheels.

This is, to us, a very satisfying solution, for having thus defined existence we proceed to speak of wheels on the train even when it is not in the station. In general, attributing a property to a system means that certain predictions about the system can be made." ("Mind, Matter, and Physicists," *Found. of Phys.*, Vol. 2, No. 2/3, 1972, pgs. 130-1.) Taken literally, this doctrine implies that the statement "Man evolved from a primate ancestor" *means* that certain bones may be seen in certain museums. This is a far from satisfying solution to us, and, we believe, to many others. It seems far more reasonable to suppose that science has simply failed insofar as it cannot deal with statements about actual reality.

29. Fine, *Theories of Probability*.

30. Kolmogorof, "Logical Basis for Information Theory and Probability Theory," *IEEE Trans.* IT-14 (1968), pgs. 662-4.

31. Chaitin, "Information Theoretic Computational Complexity," *IEEE Trans.* IT-20 (1974), pgs. 10-15.

32. Write $\lg(x)$ for the number of digits in base 32 needed to write down x. Then we have $L_m(b) \leqslant 6 \times \lg(b) \leqslant 6 \times \lg[L(Y|X,I,n)]$. Then we can replace the exponent of 2 in (45) by:

$$1 + 6 \times 39 + 6 \times \lg[L(Y|X,I,n)] - L(Y|X,I,n) \qquad (58)$$

This is more complicated, but it enables us to avoid worrying about the bound of 32^4 on $L(Y|X,I,n)$. This can also be carried over to (48) in corollary 4.

33. We should note that the function, I, may be taken to be a recursive function. To insure this, we need only require that the computer language have a criterion for rejecting programs that may fail to produce their output in a finite time. This can easily be done is a system that is suited for the calculations required by modern physics.

Bibliography

Berg, Howard C. "How Bacteria Swim." *Scientific American*, Vol. 233 No. 2, August 1975, pp. 36-44.

Chaitin, G.J. "Information Theoretic Computational Complexity." *IEEE Trans.* IT-20 (1974), pp. 10-15.

Cloud, Preston, ed., *Adventures in Earth History.* San Francisco: W.H. Freeman & Co., 1970.

Fine, Terrence L. *Theories of Probability.* New York and London: Academic Press, 1973.

Fisher, Ronald. *The Genetical Theory of Natural Selection*, 2nd ed. New York: Dover, 1958.

Freundlich, Y. "Mind, Matter, and Physicists." *Foundations of Physics*, Vol. 2, No. 2/3, 1972.

Kervran C. Louis. *Biological Transmutations.* Binghamton, New York: Swan House Publishing Co., 1972.

Kolmogorof, A.N. "Logical Baiss for Information Theory and Probability Theory." *IEEE Trans.* IT-14 (1968), pp. 662-664.

Macbeth, Norman. *Darwin Retried: An Appeal to Reason.* Boston: Gambit, 1971.

Messiah, Albert. *Quantum Mechanics.* New York: Interscience Pub., 1961-63.

Newton's Principia, Motte, Andrew, trans. New York: Daniel Adee, 1846.

Oparin, A.I. *The Origin of Life*, 2nd ed. New York: Dover, 1953.

Orgel, Leslie E. *The Origins of Life.* New York: Wiley, 1973.

Satir, Peter. "How Cilia Move." *Scientific American*, Vol. 231, No. 4, October 1974, pp. 44-52.

Tolman, Richard C. *The Principles of Statistical Mechanics.* Oxford: Clarendon Press, 1938.

von Helmholtz, H. *Über die Erhaltung der Kraft*, Ostwald's Klassiker der Exakten Wissenschaften Nr. 1, 1847.

von Neumann, John. *The Mathematical Foundations of Quantum Mechanics.* Princeton: Princeton University Press, 1955.

von Neumann, John. *Theory of Self-Reproducing Automata*, ed. Arthur Burkes. Urbana, Illinois: University of Illinois Press, 1966.

Wallace, Alfred Russell. *The World of Life.* New York: Moffat, 1911.

Watson, James D. *Molecular Biology of the Gene*, 2nd ed. New York W.A. Benjamin, 1970.

About the Author

Sadāputa dāsa (Richard Thompson) was born in Binghamton, New York, on February 4, 1947. In 1969 he earned his B.S. degree in mathematics from the State University of New York at Binghamton, and in 1970 he earned his M.A. in mathematics from Syracuse University. After receiving a National Science Fellowship in 1970, he completed his Ph.D. in mathematics at Cornell University in June of 1974, specializing in probability theory and statistical mechanics. His dissertation has been published as Memoir number 150 of the American Mathematical Society, "Equilibrium States on Thin Energy Shells."

Throughout his studies, the author was struck by the lack of any meaningful foundation to reality in modern scientific theories. His dissatisfaction with this culminated in 1970, when he studied the reduction of man to a Turing machine, a kind of abstract clockwork with a few moving parts. Surely, he felt, the truth must be something different from this. Consequently, he began to study many different philosophies, with a view to finding a practical route to higher knowledge.

In 1972 he discovered some of the books of His Divine Grace A.C. Bhaktivedanta Swami Prabhupāda in a book store in Ithaca, New York, and was struck by the beauty of their conceptions and the clarity of their presentation. Later he met the disciples of Śrīla Prabhupāda at the Rādhā-Krishna Temple in New York City. Here, he found, was a deeply meaningful philosophy capable of practical application in day-to-day life. He became formally initiated as Śrīla Prabhupāda's disciple in 1975 at the temple of Śrī Śrī Gaur-Nitai in Atlanta, Georgia. He is now a full-time member of Bhaktivedanta Institute for Higher Studies.

Monograph 3

CONSCIOUSNESS
AND THE LAWS OF NATURE

The Bhaktivedanta Institute
Monograph Series Number 3

CONSCIOUSNESS
AND THE LAWS OF NATURE

by
Richard L. Thompson
(Sadaputa Dāsa Adhikārī)

CONSCIOUSNESS
AND THE LAWS OF NATURE

The Bhaktivedanta Institute Monograph Series:

Number 1.   What is Matter and What is Life?

Number 2.   Demonstration by Information Theory that Life Cannot Arise from Matter

Number 3.   Consciousness and the Laws of Nature

Information regarding these monographs is available upon request from Bhaktivedanta Institute at:

70 Commonwealth Avenue   Hare Krishna Land
Boston, Massachusetts 02116   Juhu, Bombay 400 054
U.S.A.   India

The Bhaktivedanta Institute
Monograph Series Number 3

Consciousness and the Laws of Nature*

by

Richard Thompson, Ph.D.
(Sadāputa Dāsa Adhikārī)

*This monograph forms part of a forthcoming book, *The Origin of Life and Matter*, by Thoudam D. Singh, Michael Marchetti, and Richard Thompson.

Published by: Bhaktivedanta Institute
Boston • Bombay

iii

Readers interested in the subject matter of this monograph are invited to send correspondence to the Bhaktivedanta Institute at the following addresses.

70 Commonwealth Avenue   Hare Krishna Land
Boston, Massachusetts 02116   Juhu, Bombay 400 054
U.S.A. (617) 266-8369   India (Phone: 57-9373)

Bhaktivedanta Gurukula and
Institute for Higher Studies
Bhaktivedanta Swami Marg
Vrindavana, Mathura
India

© 1977   Bhaktivedanta Book Trust

Printed in the United States of America

Library of Congress Catalogue Card Number: 77-89118

Dedicated to His Divine Grace

A. C. Bhaktivedanta Swami Prabhupāda

om ajñāna-timirāndhasya jñānāñjana-śalākayā

cakṣur unmīlitaṁ yena tasmai śrī-gurave namaḥ

About Bhaktivedanta Institute

Bhaktivedanta Institute is a center for advanced study and research into the Vedic scientific knowledge concerning the nature of consciousness and the self. The Institute is the academic division of the International Society for Krishna Consciousness. It consists of a body of scientists and scholars who have recognized the unique value of the teachings of Krishna Consciousness brought to the West by His Divine Grace A. C. Bhaktivedanta Swami Prabhupāda. The main purpose of the Institute is to explore the implications of the Vedic knowledge as it bears on all features of human culture, and to present its findings in courses, lectures, monographs, books, and a quarterly journal, *Sa-vijñānam*.

The Institute presents modern science and other fields of knowledge in the light of Vaiṣṇava philosophy and tradition, providing a new perspective on reality quite different from that of our modern educational systems. One reason for the increasing interest of modern intellectuals in Śrīla Prabhupāda's teachings is doubtlessly the growing awareness that in spite of great scientific and technological advancements, the real goal of human life has somehow been missed. The philosophy of Bhaktivedanta Institute provides a meaningful answer to this concern by proposing that life—not matter—is the basis of the world we perceive.

The central doctrine of modern science is that all phenomena, including those of life and consciousness, can be fully explained and understood by recourse to matter alone. The dictum that "life is a manifestation of matter" is, indeed, the ultimate rationale for the entire civilization of material aggrandizement. The Vedas, on the other hand, teach that conscious life is original, fundamental, and eternal. This is the essence of *Bhagavad-gītā*—"*aham sarvasya prabhavo mattaḥ sarvam pravartate.*" (10.8) On this fundamental and critical point, modern science and Vedic knowledge find themselves opposed.

Bhaktivedanta Institute is dedicated to disseminating this most fundamental knowledge throughout the world. The Institute is clearly demonstrating that the Vedic version is not a matter simply of "faith" or "belief", but is scientific in the strict sense of the term. Although many of its features may appear difficult to verify experimentally, others have direct implications concerning what

we may expect to observe. Thus, this view should serve as a stimulating challenge to the truly scientific spirit to go beyond the very restrictive framework imposed on our scientific understanding of nature over the last two hundred years. Modern science began as an experiment to see how far nature could be explained without invoking God. But the purpose of Bhaktivedanta Institute is to introduce Vedic knowledge on a genuinely scientific basis for the first time in the history of this modern scientific age.

x

CONTENTS

I.

Introduction

> It is premature to reduce the vital process to the quite insufficiently developed conception of 19th and even 20th century chemistry and physics.
>
> —*Louis de Broglie*

At the present time it is widely claimed that life can be understood simply as a complicated interaction of atoms and molecules in accordance with known physical laws. High school and college textbooks of biology begin with the study of chemical bonding, proceed on to molecular evolution, and flatly assert that scientists have "been able to synthesize the stuff of life in a laboratory flask."[1] In scientific books and journals the theory that life is a combination of material elements is widely accepted as nearly unquestionable fact. The biochemist James Watson sums up this viewpoint as follows:

> We see not only that the laws of chemistry are sufficient for understanding protein structure, but also that they are consistent with all known hereditary phenomena.
>
> Complete certainty now exists among essentially all biochemists that the other characteristics of living organisms (for example, ... the hearing and memory processes) will all be completely understood in terms of the coordinative interactions of small and large molecules.[2]

Watson later goes on to assert that these molecular interactions have been understood by the modern theory of quantum mechanics.

> The various empirical laws about how chemical bonds are formed were put on a firm theoretical basis. It was realized that all chemical bonds, weak as well as strong, were based on electrostatic forces.[3]

In this paper we argue that this view of life is extremely shortsighted. Not only has life not been understood as a product of matter, but our understanding of matter itself is seriously deficient. We will show, in fact, that in order to remedy the deficiencies in our concept of matter we are forced to adopt an understanding of life completely different from the accepted scientific view.

We will review some of the important theories of nature of the modern scientific age. This review will culminate in a more detailed account of the present dominant theory of quantum mechanics. We shall see that none of these theories have been successful. The theories preceeding quantum mechanics have all been rejected for various reasons, and the quantum theory itself possesses serious defects which rule it out as a valid understanding of nature.

Since the time of Newton all major scientific theories of nature have been characterized by two assumptions:

(1) All of the significant features of nature can be described by numbers.
(2) All of the phenomena of nature are governed by laws which can be described by very simple mathematical equations relating these numbers to one another.

Furthermore, throughout the history of modern science, scientists have strongly tended to assume that all phenomena can be accounted for (at least in principle) by the accepted laws of their day.

These assumptions form the foundation for the modern scientific view of the absolute truth. The *absolute truth* can be defined as the ultimate causative principle or agency underlying all of the phenomena of nature; and the understanding of this fundamental cause can be seen as the goal of all fundamental research in science. However, conditions (1) and (2) impose a very severe *a priori* restriction on the nature of the absolute truth. There is no particular reason to suppose that every significant feature of nature can be described by numbers, or that those which can be so described are governed by simple equations. Our thesis is that nature cannot actually be understood within the framework imposed by these conditions.

In particular, we are proposing that the phenomenon of consciousness cannot be described by numbers, and that the behavior of matter is less and less amenable to description by simple equations the more intimately it is associated with consciousness. Since consciousness is a feature of life (at least on the human level) this thesis directly contradicts the theory that life is only a product of chemical reactions obeying simple physical laws. It is perhaps ironic, then, that compelling support for it is to be found not in the science of biology (in this paper, at least) but in physics, the fundamental study of inanimate matter.

As we shall show, consciousness is directly involved with basic problems in the quantum theory which cannot be resolved within the framework of conditions (1) and (2). Since this theory is solidly based on these assumptions, it cannot be correct as it stands. As a solution to these problems we will therefore outline a description of nature in which consciousness appears as a primary, irreducible feature of the absolute truth, lying beyond the reach of mathematical description.

In this description, consciousness enters as a natural analogue of the basic laws of nature figuring in the conventional theories of physics. It thus plays the role of an active constituent of nature. The following chart compares the role of consciousness in this view to the principle of electrical interaction in standard physics.

Entities	Electrons	"Quanta" of consciousness
Principle of interaction	Electric field	Absolute consciousness

Just as the electrons interact with other matter through the agency of the electric field (in the standard theory), so the individual conscious entities, or "quanta" of consciousness, interact with matter through the agency of absolute consciousness. However, whereas the electrical interaction is described by certain simple equations (called Maxwell's equations in the standard theory), the interaction of the conscious entities cannot be described in this way. This mathematical indescribability will be reflected by the behavior of matter—insofar as matter is affected by these interactions, its behavior will also defy reduction to any simple mathematical scheme.

Although this description of nature forms a natural extension of the standard theories, it is also consistent with a much older conception of nature epitomized by the ancient Sanskrit text, the *Bhagavad-gītā*.[4] As such it has many profound implications about the nature and potentialities of life which take us far beyond the limited schemes of modern biological theory. We feel that it deserves serious consideration both for this reason and for the elegant and natural way in which it resolves basic difficulties in modern physics.

Once when the physicist Niels Bohr was asked whether the known laws of physics would account for life, or whether life involved some higher principles as yet unknown, he replied that he did not know. He went on to say, however, that a scientist must be very conservative in his thinking, and very hesitant to discard old concepts or adopt new ones unless compelled to do so by overwhelming evidence.[5] He said that we should therefore act on the assumption that the known laws of physics, as embodied in the theory of quantum mechanics, would suffice to give us a complete understanding of the phenomena of life.

We would like to suggest that Bohr's conservatism was misplaced. Bohr was proposing that conservatism means sticking to the most radical and speculative theory that science had yet produced —a theory which was poorly understood, full of unresolved paradoxes, and tested only in limited circumstances.

A much more fruitful conservative approach entails the ancient understanding that the fundamental principle of life is an entity— the self or *atma* (quantum of consciousness)—which is not reducible to matter. As such, this approach is not merely a theoretical exercise, but it has many practical, empirical consequences. In particular, it entails the direct observation and study of the *atma* and its relationship with the absolute consciousness, or *paramatma*. Some of the principles governing this study are briefly outlined in the last section.

4

II.

The Representation of Nature by Mathematical Laws

Historical Outline

The wise have explained that one result is derived from the culture of knowledge and that a different result is obtained from the culture of nescience.

—*Śrī Īśopaniṣad*

The earliest known attempts to describe nature by means of mathematical laws date back to the most ancient of times and concern the subject of astronomy. These involve tables of numbers and systems of calculation used to predict the positions of the sun, moon, and planets at different times. Also dating back to ancient times were the Pythagorean philosophers who propounded the doctrine that everything in nature is based on numbers and acts according to numerical laws. They are known for their description of the musical tones produced by vibrating strings in terms of numerical ratios.

The speculations of these philosophers demonstrated some of the characteristic features which, as we shall see, have been shared by all attempts to describe nature by mathematics. One of these features is the tendency to regard some mathematical structure or relationship as an absolute principle underlying all reality. Thus the Pythagoreans thought that reality must be based on the properties of ratios of whole numbers. Likewise, certain Greek astronomers believed that the circle must be the basis of all motion, and they therefore attempted to describe the motions of the planets in terms of a system of wheels turning on wheels. Both of these schemes also exhibited another characteristic feature. They were both eventually seen to be false as fundamental principles and of limited value as descriptive devices. Neither is given much importance today.

In more recent times Galileo studied the flight of heavy objects through the air and found that their trajectories closely approximated the geometrical curves known as parabolas. With the development of analytic geometry by Descartes, it became possible to represent such curves by numbers, and by simple equations relating these numbers to one another. This is illustrated in Figure

5

1. The vertical and horizontal displacements of a cannonball at times t_1, t_2, and t_3 are represented by vertical coordinates y_1, y_2, and y_3 and horizontal coordinates x_1, x_2, and x_3. In this example the x and y coordinates for the curve representing the flight of the cannonball are related by a simple formula: $y = 100 - x^2/10$.

With the work of Newton this kind of representation of nature was greatly extended. Newton developed general laws for computing the trajectories of flying objects, and he applied these both to heavenly bodies and objects on the earth. His success in this attempt marked the beginning of the modern period in the development of science, in which the ideal goal of scientific investigation has been to describe nature in terms of exact numerical relationships between measurable quantities.

Newton dealt mainly with mathematically predicting the movements of large objects, such as cannonballs or planets, but he made the suggestion that his basic principles might have universal applicability.[6] As scientists began to develop and extend his work, the conception became widely accepted that all of nature could be described by numbers corresponding to measurable quantities, and that all the phenomena of nature could be described by mathematical "laws" relating these numbers to one another. Thus, not only would the gross movement of a cannonball be numerically described, but the cannonball itself, the cannon, the man who fired it, and ultimately the entire universe.

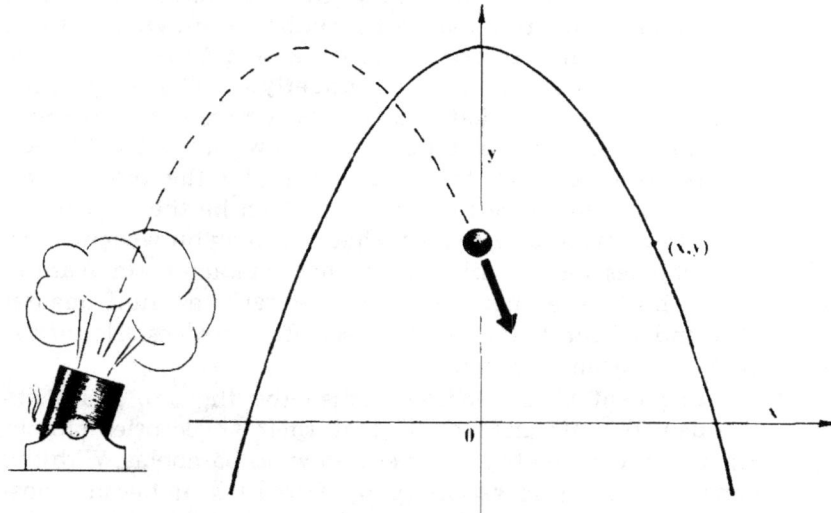

Figure 1. Galileo's law for the flight of a cannon ball.

This program required the drastic philosophical assumption that there exists an exact one-to-one correlation between the features of reality and numbers in some mathematical system. In fact, this assumption, which is reminiscent of the philosophy of the Pythagoreans, soon became almost universally accepted among Western scientists. Here is a recent formulation of this assumption by Albert Einstein.

> Every element of the physical reality must have a counterpart in the physical theory. We shall call this the condition of completeness.[7]

Here the term "physical theory" refers to some mathematical system, and "physical reality" simply means reality, for everything is assumed to be physical.

By the beginning of the nineteenth century it was believed that this program had in fact been successfully carried out, and thus the famous French scientist Pierre de Laplace was able to assert that "all the effects of nature are only the mathematical consequences of a small number of immutable laws."[8] He believed that the universe is made up of atomic particles and that the exact condition of the universe at any one time could be given by specifying the exact positions and velocities of those particles with respect to a system of coordinates. He claimed that given these positions and velocities he could, at least in principle, calculate the entire past, present, and future of the universe from the "laws of motion" governing the particles.

This is the famous mechanical conception of the universe which became the foundation of the nineteenth century philosophy of materialism. Essentially, the mathematical equations Newton had used to describe the gross movement of large objects were applied to atoms and interpreted as the universal principles underlying the world as a whole. This world view is illustrated in Figure 2. The atoms are depicted as little balls moving through space under the influence of forces of mutual attraction and repulsion. Each atom is described by seven numbers: a mass, m, three coordinates of position, q_1, q_2, and q_3, and three coordinates of momentum, p_1, p_2, and p_3. The complete list of these numbers for all the atoms making up the universe was taken as a complete and exact numerical description of reality in total. Thus the atom was accepted as the ultimate "element of physical reality" in the sense of Einstein's definition, and was not regarded as having any features apart from position, momentum, and mass.

7

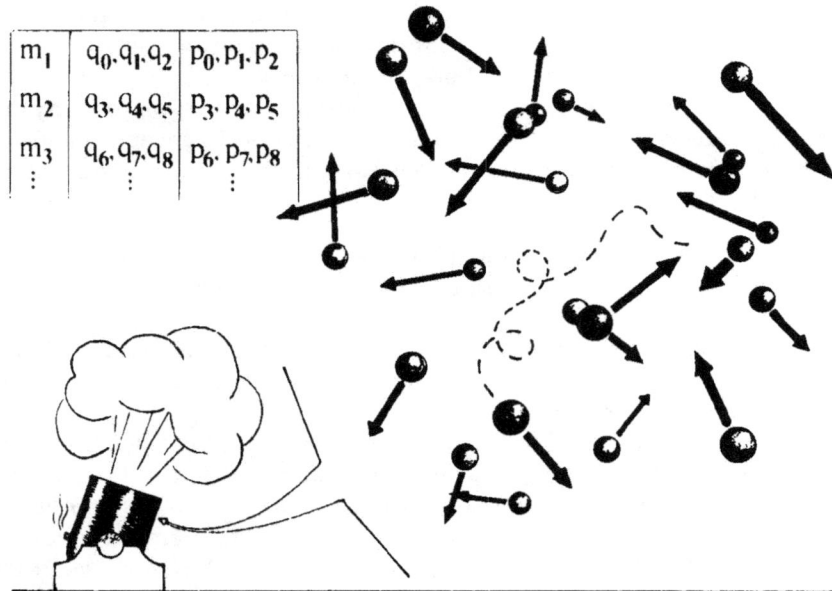

Figure 2.   The Laplacian view of the world
            as a swarm of atomic particles.

In Figure 3 we have indicated the equations governing the motion of these atoms. These are Newton's laws in a form developed by the mathematician William Hamilton.

We have described this picture of reality in some detail because it possesses the essential features of all subsequent scientific descriptions of nature. In this system the exact motion of all the atoms is determined by equations (1) through (4) and their positions and velocities at one particular moment. These equations are thus referred to as the "laws of nature." The positions and velocities may be specified for any moment in time, but since they are generally chosen for the beginning time of the phenomena being investigated, they are called initial conditions. Taken together, these natural laws and the initial conditions constitute the absolute truth in this world system, since no further causative principle is granted to exist.

In this world view, the absolute truth appears as a rather awkward composite of two quite dissimilar things. The initial conditions are simply an arbitrary list of data, whereas the natural laws appear as ultimate, all-pervading, and non-material causative principles. Fundamental natural laws are presumed by definition

$$\frac{dq_k}{dt} = \frac{\partial H(\overline{p},\overline{q})}{\partial p_k} \quad (1) \qquad \frac{dp_k}{dt} = -\frac{\partial H(\overline{p},\overline{q})}{\partial q_k} \qquad (2)$$

$$H(\overline{p},\overline{q}) = \sum_i \frac{p_{3i}^2 + p_{3i+1}^2 + p_{3i+2}^2}{2m_i} + V(\overline{q}) \qquad (3)$$

The potential, V, represents the forces of attraction and repulsion between the atoms, and is of the form:

$$V(\overline{q}) = \sum_{i>j} \frac{K_{ij}}{\sqrt{\sum_{k=o}^{2} (q_{3i+k} - q_{3j+k})^2}} \qquad (4)$$

Figure 3. Newton's laws.

to be invariant both in time and space. If a law is to change from moment to moment, or from place to place, then it is normally assumed in scientific practice that this change must be due to the influence of an underlying law which *is* spatially and temporally invariant.

Also, the laws of nature must be due to some existing but non-material cause—something that is real but that cannot be visualized in our customary way as a combination of some kind of elements. To illustrate this, consider two cannonballs separated by miles of empty space. According to the law of gravity, they will accelerate toward one another. What is it that acts across empty space to pull each ball unerringly toward the other? If the law of gravity is fundamental then we cannot break down this "something" into a combination of other things. If not, then other more basic laws are presumably involved. As we shall see, all the fundamental laws considered in the history of modern physics—such as Coulomb's law and Maxwell's laws of electromagnetism—have the same mysterious character. Something real must be acting, but that something is essentially inconceivable to the human mind.

It is interesting to note that this very feature of Newton's system was rejected as "mysticism" and "occultism" during his time by the mechanistic school of natural philosophers in Europe. These men wanted to reduce all phenomena to direct physical pushes of object against object, but had failed to accomplish this.

9

However, even if they had succeeded, they would still have had to introduce some sort of inconceivable fundamental law.[9] Newton himself maintained that "it is inconceivable, that inanimate brute matter should, without the mediation of something else which is not material, operate upon and affect other matter without mutual contact."[10]

The initial conditions are a rather unsatisfying feature of this system because no indication is given as to why one particular set of initial positions and velocities should be chosen and not another. Newton had supposed that the initial conditions were chosen by God, Who then retired from the scene and allowed the universe to run automatically by Newton's laws. Other scientists, such as Laplace, were adamantly opposed to admitting the existence of God in any form, and they were forced to propose that the initial conditions just arbitrarily happened to be the way they were.

This idea was rendered somewhat less unpalatable by supposing that initially (at some time in the past) the atoms were arranged in the form of a big cloud of gas—a primordial nebula. It was supposed that Newton's laws would then act to generate the sun, the earth, the seas and continents, and the various living organisms, and that the atoms in the primordial cloud did not have to be arranged in any particularly precise way in order for this development to take place. Of course, this still left the origin of the cloud unexplained. It also elevated the laws of nature even more thoroughly into the position of the ultimate cause of all causes by granting them the capacity to create the entire universe out of chaos.

This ultimate cause of all causes, however, was regarded as being equivalent to nothing more than two or three lines of equations! In this system it was proposed that *everything* could be explained in terms of the composite action of large numbers of simple parts interacting according to extremely simple rules. The idea of describing nature in terms of combinations of atoms was not in itself new. The ancient Indian philosophers Gautama and Kanāda had taught that "atomic combination is the original cause of the creation,"[11] and the Greek philosopher, Democritus, had declared that, "ostensibly there is color, ostensibly sweetness, ostensibly bitterness, actually only atoms and the void."[12] The distinguishing characteristic of the modern systems has been the idea of completely describing the atoms and their behavior by means of the smallest possible number of exact mathematical laws.

This world view thus reduced man to nothing but a transient

combination of atoms, each interacting with the others by simple pushes and pulls, and it therefore could give no account of the existence of consciousness. Since conscious awareness is the most immediate and undeniable feature of all of us, this point is very significant and deserves our careful attention. The physicist Eugene Wigner called consciousness "the first kind of reality" and noted that "the reality of my perceptions, sensations and consciousness is immediate and absolute."[13] All of our knowledge of the world is in fact dependent on this reality. Yet, no hint of it is to be found in the world picture depicted in Figure 2.

This deficiency has generally been covered over by the confusion of consciousness with behavior. It is alleged that the interaction of many atoms may indeed add up to produce the very complex patterns of motion characteristic of say, a human being. However, even if this were true, this complicated atomic motion could only account for the patterns of sensations, thoughts, and feelings. It could not account for our awareness of these things.

After all, if there exists nothing but atoms rushing through space, and each atom is featureless and insentient, then the fact that these atoms happen to form a certain geometrical pattern on a large scale gives us no conception at all of why there should exist conscious awareness of that pattern. Each atom is a separate entity characterized by a particular momentum, mass, and position, and the total system is just a collection of these separate atoms. Why should there exist conscious awareness if atoms in the retina of an eye are arranged in the form of a certain image, or if the atoms in the brain are arranged in another pattern bearing some definite relationship with this one? Imagine a vast cloud of atoms similar to that depicted in Figure 2; if this is indeed all that there is, then who or what is perceiving the appearances of taste and color mentioned by Democritus?

It might be objected here that perhaps the atoms possess other features that enable them to generate consciousness by combining together. The answer is that in this case we would be dealing with another theory. The fundamental elements in this theoretical system have already been identified, and no other features of this sort are included among them. If a mathematical theory of these features were possible, then they would have to correspond to arrangements of numbers within that theory. However, as we examine each of the main scientific theories of nature in turn, we shall not find such a numerical scheme. An alternative possibility is that consciousness may correspond to features of reality which are not amenable to mathematical description. We propose that this is in fact the case, and we shall consider this in greater detail later.

Another important feature of this world system is that it is not at all clear that matter actually does behave according to the laws of motion. It is well known that Newton's laws of motion can be solved exactly for at most two moving bodies if we assume that forces vary with the inverse square of the distance. (This is the standard assumption.) For larger numbers of bodies approximations must be used, and these quickly become very involved and cumbersome as the number increases. For systems of many billions of atoms there is no question of even approximately solving the equations of motion, and therefore there is no way of practically finding out whether the theory really does duplicate the phenomena of nature. In view of its serious deficiencies, it therefore seems astonishing that this system could be seriously proposed as an exact description of universal reality.

Of course, the precise model of reality proposed by Laplace is no longer considered to be valid. In the 1860's a revolutionary change took place in the scientific conception of the fundamental nature of matter. An entirely new type of mathematical construct was introduced into physics by James Clerk Maxwell in order to account for the phenomena of electricity and magnetism, and this was accompanied by a new system of natural laws. Maxwell's theory was considered to be particularly important because it was able to account for the wave-like properties of light discovered by other investigators, such as Fresnel and Young. These properties had never been successfully explained by the earlier theory.

As a result of this revolution, physicists now employed vector fields in addition to the atomic particle coordinates considered by Laplace. A vector may be thought of as an arrow pointing in a certain direction with its base at a certain location. A vector field may be thought of as an array of such arrows arranged so that each point in space lies at the base of exactly one of the arrows. In Maxwell's theory of electromagnetism, two vector fields of this type were introduced—the field, $\bar{E}$, to describe electrical forces, and the field, $\bar{H}$, to describe magnetic forces.

Figure 4 illustrates a typical arrangement of these two fields. The horizontal arrows represent the magnetic field, and the vertical arrows represent the electrical field. The complete distribution of these arrows in the form of perpendicular sinosoidal curves is shown along the x-axis. The arrows may be visualized in a similar

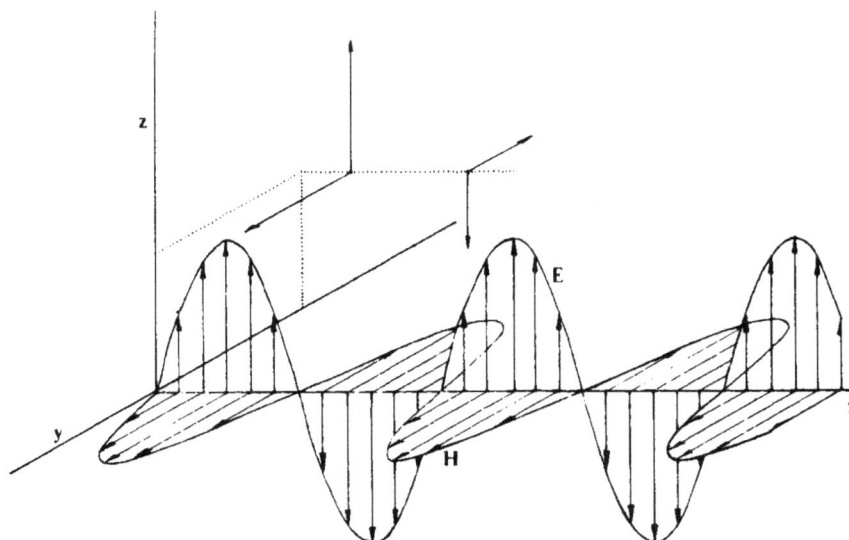

Figure 4. The vector fields for a beam of polarized light.

distribution along each line parallel to the x-axis. This picture represents a beam of light of a single frequency moving in the x-direction, as described by Maxwell's theory.

Figure 5 depicts the equations introduced by Maxwell to describe the transformation of the electric and magnetic fields with the passage of time. Here, ρ refers to the density of charged particles in space and $\bar{J}$ describes their flow. $\bar{J}$ is called electrical current. In this theory the atomic particles were given the new property of electrical charge, which could be positive, negative, or neutral.

For some time after this new theory was introduced, scientists attempted to represent the vector fields somehow or other in terms of arrangements of moving atoms and other mechanical con-

$$\nabla \times \bar{E} = -\ \frac{1}{c}\ \frac{\partial \bar{H}}{\partial t} \quad (5) \qquad \nabla \times \bar{H} = -\ \frac{1}{c}\ \frac{\partial \bar{E}}{\partial t}\ + \quad (7)$$
$$4\pi \bar{J}$$

$$\nabla \cdot \bar{H} = O \qquad (6) \qquad \nabla \cdot \bar{E} = 4\pi\rho \qquad (8)$$

Figure 5. Maxwell's equations.

13

structs, such as fluids and elastic solids, which could be treated as limiting cases of systems of atoms. In this way they tried to retain the absolute and final status they had attributed to their original theory. Gradually, however, they abandoned these attempts as unworkable and adopted the view that electric and magnetic fields have an existence of their own which is independent of atoms. The world was now visualized as consisting of two components: vector fields extending continuously through space, and large numbers of localized atomic particles. The original natural laws indicated in Figure 3 were now amended to include Maxwell's equations, and terms were also added to them corresponding to the forces exerted on the particles by the fields (the Lorentz force law).

Apparently some thirty years were required for this transition in thinking to become accepted. The physicist Freeman Dyson has given an interesting explanation as to why so much time was needed for this.[14] According to him, the scientists of that time were limited in their thinking to the concepts and language of their old theory, and as a result they were practically unable to conceive of anything new. Max Planck has also described this phenomenon by remarking that "A new scientific truth does not triumph by convincing its opponents and making them see the light, but rather because its opponents eventually die, and a new generation grows up that is familiar with it."[15] We can understand from this that even though a given scientific theory falls far short of giving a complete and correct account of reality, those people who are limited in their education to the concepts and language of that theory will be unable to see this without great difficulty.

This phenomenon of human psychology has serious implications concerning the scientific process of acquiring knowledge. As we will see later on, there is ample reason to think that the current generation of scientists and students of science is similarly bound within a limited and defective conceptual framework. The slow and painful procedure of gradually devising very limited concepts, falling under the illusion that they represent universal truth, and then yielding to new concepts, can only be expected to lead to true understanding if reality is a very limited and simple affair. If

$$M \frac{d^2 \bar{q}}{dt^2} = e \left[\bar{E} + \frac{1}{c} \frac{d\bar{q}}{dt} \times \bar{H} \right] \tag{9}$$

Figure 6. The Lorentz force law.

it is not so simple, then this procedure can only be expected to lead to practically permanent entanglement in illusion and ignorance. In the last section we will outline an alternative method of obtaining knowledge which is opened up if reality is not, in fact, limited to the narrow scheme of numerical representability which has been adopted by modern science. For now, however, we shall continue to examine the predicament of modern attempts to devise a mathematical description of nature.

As soon as the transition to the new ideas was accomplished, the new world system was given the same complete and absolute status that had once belonged to the earlier theory. Some scientists in England even went so far as to declare that now everything fundamental about reality had been discovered and that scientists of the future would have nothing left to do but work out the consequences of these discoveries in greater and greater detail. This in fact became the prevailing view among physicists toward the end of the nineteenth century.

The same fundamental considerations we made concerning the Laplacian atomic theory also apply with equal weight to this world view. In particular, the addition of vector fields to the physical picture cannot account for the phenomenon of conscious awareness. Suppose that we now visualize a brain as an immense swarm of atomic particles situated in the midst of a great sea of arrows extending to various lengths in various directions. If these move according to the natural laws, then each will vary in position and direction in a simple and rigidly determined way depending on the arrangement of the other arrows and particles. Since there is nothing there but arrows and particles in motion, we have no reason to suppose that there will exist any awareness there, regardless of what the overall arrangement of the arrows and particles may be. The fundamental ingredient that yields consciousness has evidently been omitted from this picture.

We should stress again that consciousness refers to the undeniably existent awareness that "I see," or "I think," as opposed to patterns of behavior, such as the vibration of the sound, "I see," which may accompany such awareness. It is sometimes claimed that consciousness will be described as a property of matter analogous, for example, to wetness or hardness. However, strictly speaking, these properties presuppose consciousness since they are perceptions. Thus, wetness is more than simply a designation for a class of material configurations which one might hope to define by some numerical relationship. Any account of wetness must include

15

the senses, the mind, and ultimately the consciousness of a perceiving person. It explains nothing to say that consciousness is analogous to a physical property which, in its very definition, makes implicit reference to consciousness itself.

It has sometimes even been proposed that consciousness is an illusion, but then who or what is deluded? We might hope to describe patterns of matter corresponding to states of wetness or states of delusion, but to account for *awareness* of these patterns is a completely different problem.

Our basic theme has been that consciousness cannot be understood simply in terms of a collection of non-sentient entities, such as arrows or particles, which are thought to constitute matter. In this system, however, we encounter a new problem which, as we shall see, has become even more serious with the development of quantum mechanics in the twentieth century: it is very hard to identify just which elements of the theoretical system are to be regarded as real and which ones should be regarded simply as mathematical artifices. This results in a complete breakdown in the idea that a mathematical description of nature can be found in which there exists a one-to-one correspondence between arrangements of numbers and "elements of reality."

In order to illustrate this, let us consider the electric and magnetic fields, $\bar{E}$ and $\bar{H}$, of the electromagnetic theory. If these are visualized as in Figure 4, we may tend to suppose that these arrays of arrows in space are to be regarded as actually existing components of reality, just as the atoms in Figure 2 were regarded as actually existent particles. However, closer examination shows that this naive view is not really justified. For example, both $\bar{E}$ and $\bar{H}$ can be expressed in terms of a new vector field, $\bar{A}$, called the vector potential. All of the various equations of the theory can then be reformulated in terms of $\bar{A}$ without any mention of the original $\bar{E}$ and $\bar{H}$. We are therefore led to inquire: is $\bar{A}$ just a mathematical artifice, or does it actually exist, and are $\bar{E}$ and $\bar{H}$ just artificial constructs?

Actually, this system can be reformulated in many different ways. In Figure 7 we illustrate a reformulation which proved to be important in the later development of quantum mechanics. In this formulation, we still find the familiar atomic particles. However, the vector fields have completely disappeared. In their place we find a new class of entities, called "radiation oscillators," which possess position and momentum but which cannot be thought of as particles having an actual existence in space. These entities are

16

q_1, p_1
q_2, p_2
q_3, p_3
$\vdots$

$m_1, e_1 \quad \bar{Q}_1, \bar{P}_1$
$m_2, e_2 \quad \bar{Q}_2, \bar{P}_2$
$m_3, e_3 \quad \bar{Q}_3, \bar{P}_3$
$\vdots \qquad \vdots$

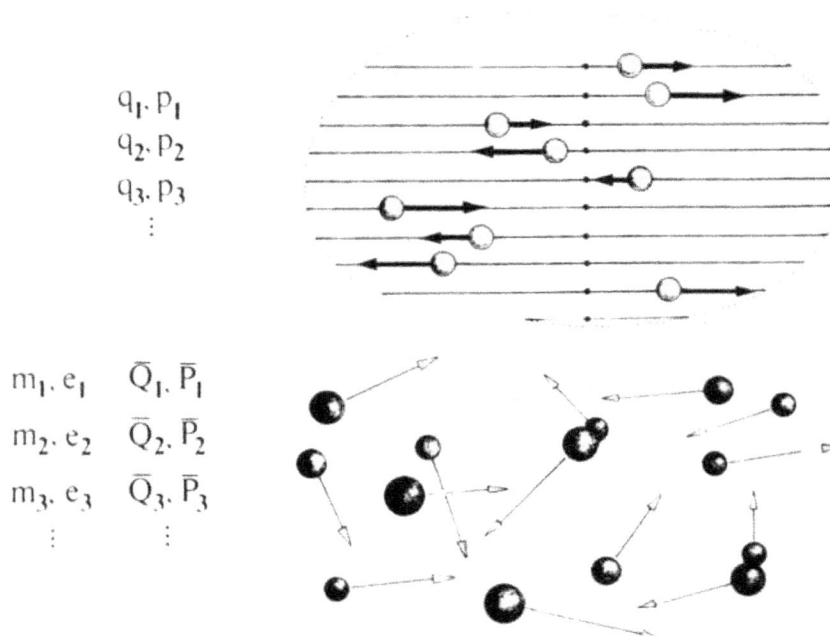

Figure 7. The "mechanical" model of atomic particles in an electromagnetic field.

governed by a new set of natural laws, illustrated in Figure 8. Since these laws possess a formal structure similar to that given for Newton's laws in Figure 3, this formulation of the theory is called the "mechanical model."[16]

The important point here is that with these developments the idea of giving an exact mathematical description of reality has completely dissolved. All of these different formulations give the same experimental predictions, because they are simply mathematical transformations of one another. Therefore, each one might just as well be regarded as a description of reality as the other. Each one might likewise be regarded as an artificial mathematical scheme. From this point on scientists have increasingly had to renounce the idea of determining an actual breakdown of reality into a combination of basic mathematically describable elements. Instead, they have been forced more and more to adopt the point of view that a mathematical theory is simply a calculational device which is of value to the extent that it correctly predicts the relationship observed between experimental measurements.

17

$$\frac{d\widetilde{q}_k}{dt} = \frac{\partial H}{\partial \widetilde{p}_k} \qquad (10) \qquad\qquad \frac{d\widetilde{p}_k}{dt} = -\frac{\partial H}{\partial \widetilde{q}_k} \qquad (11)$$

$$\frac{d\bar{Q}_k}{dt} = \frac{\partial H}{\partial \bar{P}_k} \qquad\qquad\qquad \frac{d\bar{P}_k}{dt} = -\frac{\partial H}{\partial \bar{Q}_k}$$

$$H(\bar{P}_i, \bar{Q}_i, \widetilde{p}_i, \widetilde{q}_i) = \qquad\qquad\qquad\qquad\qquad\qquad (12)$$

$$\sum_i \frac{c^2 \widetilde{p}_i^2 + \eta_i^2 \widetilde{q}_i^2}{2} + \sum_i \frac{(\bar{P} - \frac{e_i}{c} \Sigma_n \widetilde{q}_n \bar{A}_n(\bar{Q}_n))^2}{2M_i} + V$$

Here, V is the same as in Figure 3, except that now electrical charge is considered. Thus $K_{ij} = e_i e_j - GM_i M_j$.

$$V = \sum_{i>j} \frac{K_{ij}}{|\bar{Q}_i - \bar{Q}_j|} \qquad\qquad\qquad\qquad\qquad (13)$$

Figure 8. Equations for the mechanical model [17] of electromagnetic theory.

This should illustrate the futility of trying to ascertain absolute knowledge by this kind of speculative approach. Even if we now wish to think that Figure 8 depicts the absolute truth, rather than Figure 3, we find that we cannot ascertain any clear idea of how these equations actually relate to reality. Returning to the question of consciousness, we can see that if we regard the elements of any one of these formulations as equivalent to the "elements of reality," then we are again left with the problem of understanding why a collection of separate insentient entities should be conscious. On the other hand, if we regard the theory simply as a computational device, then we are admitting that it is an incomplete description of reality which need not even be expected to refer to consciousness. It cannot be thought of as complete or absolute.

However, this theory did not retain its absolute status for long. It was soon realized that it failed in many ways to give correct descriptions of the phenomena of nature. One of the most dramatic of these failures lay in its prediction that matter would be unstable. According to the theory, matter must be composed of nearly equal numbers of positive and negative charges since it tends to be electrically neutral on a large scale. However, the theory predicted that a positive and a negative particle would tend to fall together and release their mutual potential energy to the radiation field. Thus, a piece of matter would be expected to quickly release a large amount of radiant energy and either explode or transform itself into a mass of electrically neutral particles. (Some assumptions about the structure of the particles were also needed here—two point charges would release an infinite amount of radiation as they fell together.)

In addition, the theory was found to be at variance with the results of an increasing number of experiments. As this evidence was accumulated, the need was gradually felt for a completely new approach. By the twenties of this century a new theory had been developed which involved a fundamental revision in the scientific concept of the nature of matter. This came to be known as the theory of quantum mechanics.

At the present time, quantum mechanics is the dominant theory of physics.[18] It is now widely believed that, at least in its basic structure, this theory provides a complete and fundamental description of reality. To be sure, many mysterious things have been discovered in the investigation of matter at high energies, and some scientists have proposed that an entirely new theory will be required to account for them. However, the general view is that these discoveries can be explained within the basic framework of the theory by the addition of new quantum laws and variables.

The basic principles of the theory are almost universally regarded as inviolable. In the range of energies in which ordinary chemical reactions take place, it is accepted that this theory gives a complete and correct account of all phenomena as it stands. As we pointed out in the introduction, life is regarded by orthodox scientists as a chemical phenomenon of this type.

As we shall see in the next section, however, this theory suffers from fundamental defects which rule it out as a valid description of matter, even on the level of ordinary chemistry.

19

III.

The Theory of Quantum Mechanics

One is no longer able to describe or even to think about any well-defined connections between the phenomena at a given time and those at an earlier time.

—David Bohm

In this section we will give a brief description of the view of nature presented by the theory of quantum mechanics. Strictly speaking, if quantum mechanics could be said to give an absolute description of the world at all, this would have to be provided by that form of the theory known as relativistic quantum mechanics. This version of the theory takes into account the theory of relativity enunciated by Albert Einstein. However, the mathematical analysis which it entails is very involved, and, in the words of one physicist, is "exempt neither of difficulties nor even of contradictions."[19] For this reason, we will deal mainly with the nonrelativistic theory (known simply as quantum mechanics or quantum theory). All the features of this theory with which we shall be interested also carry over to the relativistic theory.

First, let us consider the basic mathematical structures used to describe nature in quantum mechanics. One of the main features of this theory is that it builds in a systematic way on the mechanical theory of the 19th century. Formally, quantum mechanics is based on a set of mathematical operations, called correspondence rules, whereby one can convert a mechanical model of nature in terms of moving particles into a new mathematical form which could be called a quantum mechanical model. This is why the mechanical model of the electromagnetic theory mentioned at the end of the last section proved useful, even though the theory was not originally visualized in that form. Since this model involved particles moving according to mechanical laws of motion, it was possible to convert it into a quantum mechanical model.

We will briefly outline the most basic elements of the theory without attempting to explain how scientists were historically motivated to introduce them. Then we will illustrate the theory by some practical examples in order to clearly bring out its implications.

The Wave Function

The basic constituent of a quantum mechanical model is an entity called the *state* of the system. The state can be represented mathematically in many different ways, but the simplest to describe is the representation known as the wave function. If the model is based on a mechanical model with particles at locations, $\bar{q}_1, \ldots, \bar{q}_n$, then the wave function will be a complex valued function, $\psi(\bar{q}_1, \ldots, \bar{q}_n, t)$, of these variables and the time, t.[20] The basic premise of the theory is that at any time, t, the wave function provides the most complete description possible of the physical system at that moment in time. Thus, the wave function plays a role analogous to the list of particle coordinates and momenta, $\bar{q}_1, \ldots, \bar{q}_n; \bar{p}_1, \ldots, \bar{p}_n$, believed to provide a complete description of the condition of the system in the 19th century mechanical theory.

The Schrödinger Equation

As time passes, the wave function is assumed to change in accordance with a certain differential equation, called the Schrödinger equation. The general form of this equation is illustrated in Figure 9(14). For any particular quantum mechanical model a specific Schrödinger equation is built up in a systematic way from the equations of motion of the corresponding mechanical model. Figure 9(15) shows the form taken by this equation when the mechanical model is the Laplacian world system illustrated in Figures 2 and 3 of the last section.

The Schrödinger equation corresponds to the basic equations of motion of the earlier theories. It determines the transformations of the wave function with time corresponding to the action of the various physical forces within the system. The quantum mechanical force laws are generally based directly on the force laws of the underlying mechanical system, although some of them involve variables, such as those describing spin, which belong uniquely to the quantum mechanical theory. All of the quantum mechanical force laws are summed up in the term, $\mathcal{H}$, which is known as the Hamiltonian operator of the system.

As in the older theories, this transformation of the wave function with time is uniquely determined by the equation of motion—the Schrödinger equation—and the initial condition, $\psi(\bar{q}_1, \ldots, \bar{q}_n, t_0)$, of the wave function at some arbitrary time, t_0, in the past. In

21

$$\mathcal{H}\psi \;=\; i\hbar\;\frac{\partial\psi}{\partial t} \tag{14}$$

$$-\hbar^2\;\sum_i\;\frac{\dfrac{\partial^2\psi}{\partial q_{3i}^2} + \dfrac{\partial^2\psi}{\partial q_{3i+1}^2} + \dfrac{\partial^2\psi}{\partial q_{3i+2}^2} \;+}{2m_i} \tag{15}$$

$$V(\overline{q})\,\psi \;=\; i\hbar\frac{\partial\psi}{\partial t}$$

In (15) the term for $\mathcal{H}$ is obtained from the H of Figure 3 by applying the substitutions, $\qquad$ (16)

$$p \rightarrow -i\hbar\;\frac{\partial}{\partial q} \text{ and } V(\overline{q}) \rightarrow \text{multiplication by } V(\overline{q}).$$

These are called correspondence rules.

Figure 9. The Schrödinger equation.

this theory also, no causative principles are granted to exist in nature other than the initial conditions and the laws of motion. However, the theory does introduce another principle of transformation—pure or causeless chance—not found in any of the earlier theories.

The Connection Between the Wave Function and Reality

In order to understand the significance of this formal apparatus we must see how the wave function, ψ, is related in this theory with the actual phenomena it is supposed to describe. Unfortunately, this relationship is expressed in terms of a very complicated and abstract mathematical scheme involving such things as Hilbert space, inner products, unbounded Hermetian operators, and systems of orthonormal basis vectors. In this paper, however, we will try to give a simple presentation and avoid introducing these complexities as much as possible. We will begin with a simple example which will indicate some of the main features of the wave function.

Figure 10 depicts a simple mechanical model and its corres-

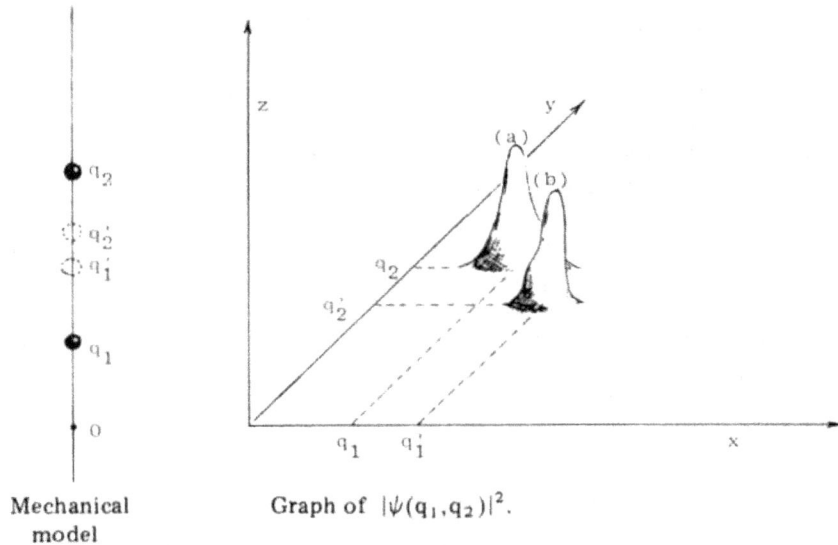

Mechanical model

Graph of $|\psi(q_1, q_2)|^2$.

Figure 10. A simple quantum mechanical model.

ponding wave function. This model can be thought of as representing two small particles which are free to move up and down on a vertical wire. The variables q_1 and q_2 measure the position of these particles on the wire, and the wave function at a fixed time will be a function, $\psi(q_1, q_2)$, of these two variables. Each point in the plane can be associated with a configuration, (q_1, q_2), of the two particles if we let q_1 and q_2 designate the x and y coordinates of the point. The absolute value of the wave function squared, or $|\psi(q_1, q_2)|^2$, may be visualized as a surface with hills and valleys standing above the plane—the height of the surface above the point (q_1, q_2) equals the value of $|\psi(q_1, q_1)|^2$. In this example, $|\psi|^2$ is assumed to be very small except in two places, marked (a) and (b), where it takes on large values.

The basic interpretation of the wave function is that its absolute value squared, $|\psi(q_1, q_2)|^2$, represents the *probability* of finding the first particle at q_1 and the second particle at q_2.[21] This is known as the Born interpretation. In this particular illustration, the wave function is shown with large values in the region (a), and this must therefore be interpreted to mean that q_1 and q_2 will be very likely to have positions corresponding to points (q_1, q_2) in this region. These are roughly the positions, q_1 and q_2, of the particles as shown in the figure.

However, the wave function is also shown to have large values

23

in the region (b). The points, (q'_1, q'_2), in this region correspond to the alternate positions of the particles shown in the figure by dotted circles. Thus we can see that the quantum mechanical wave function does not give an exact description of nature—it is by definition statistical. It is consequently possible for the wave function to give an ambiguous account of the arrangement of matter it describes.

In this example, the wave function indicated two alternative arrangements for the particles and did not distinguish between them. In order for the function to indicate an exact location for the two particles it would have to be concentrated on one point, (q_1, q_2), and be equal to zero at all other points. On the other hand, if the function is non-zero over the whole xy plane, then it may not give any indication at all about the locations of the particles.

These traits of the wave function also apply to general mechanical systems, although they become harder to visualize since the wave functions for systems of many particles have their domains in highly multi-dimensional space. For a system of n particles the wave function will have 3n variables, since the location of each particle in three dimensional space is given by three coordinates. For such a function, the diagram in Figure 10 would correspond to a 3n dimensional "surface" in 3n + 1 dimensional space. For large systems, such as a model of a human being, this number of dimensions will be enormous. As an indication of the size of n, consider that one gram molecular weight of a substance (e.g., 32 grams of oxygen, O_2) is estimated to contain 6×10^{23} molecules, each involving several particles, such as electrons and nuclei. This shows that the wave function cannot in any way be thought of as an entity actually existing in three dimensional space in the way the particles and electromagnetic fields of the earlier theories were believed to exist.[22]

Such a wave function can also be highly ambiguous in its description of matter. Figure 21 indicates the kind of ambiguous account which can be given by the wave function for a theoretical model of a human being. The possibility of such ambiguity becomes especially striking if we consider that in the theory of quantum mechanics, the wave function is considered to provide a *complete* description of the material arrangement of the system. In the words of one prominent physicist, "given any object, all the possible knowledge concerning that object can be given as its wave function."[23] No more extensive knowledge about the system is considered to be possible, even in principle. This means that if the system is in a state corresponding to an ambiguous wave function, then the ambiguity must be considered to be absolute—the material system actually *is* ambiguous. In the example in Figure 21, this implies that the man is neither in one situation nor another—his situation is *essentially indeterminant*.

The Heisenberg Uncertainty Principle

Before describing how this rather contradictory looking situation is dealt with in the theory of quantum mechanics, we should first note that *every* quantum mechanical wave function must possess some degree of ambiguity.

Thus far we have indicated the relation posited between ψ and the positions of particles in the underlying mechanical system. In addition to the particle coordinates, $\bar{q}_1, \ldots, \bar{q}_n$, the classical theory also dealt with the momenta, $\bar{p}_1, \ldots, \bar{p}_n$, of the particles. (Momentum is defined as mass times velocity, or $\bar{p} = m\bar{v}$.) According to the theory, there is a transformation which will convert $\psi(\bar{q}_1, \ldots, \bar{q}_n)$ into a new function, $\phi(\bar{p}_1, \ldots, \bar{p}_n)$, which depends on the momentum coordinates. This transformation, known technically as a Fourier transform, is illustrated in Figure 11. This form of the wave function enables us to find the momenta of the particles. Just as $|\psi(\bar{q}_1, \ldots, \bar{q}_n)|^2$ was interpreted as representing the probability that the particles will be found in the locations, $\bar{q}_1, \ldots, \bar{q}_n$, so the value of $|\phi(\bar{p}_1, \ldots, \bar{p}_n)|^2$ is interpreted as representing the probability that the particles have momenta, $\bar{p}_1, \ldots, \bar{p}_n$

The transformation between ψ and ϕ is such that there must be a reciprocal relation between the ambiguity of ψ and that of ϕ. This relationship is summed up in the famous Heisenberg uncertainty principle,

$$\Delta P \Delta Q \geqslant \hbar/2$$

Here, ΔP measures the ambiguity in the momentum of some part of the system (i.e., of some particular collection of particles), and ΔQ measures the ambiguity in its position. The quantity, $\hbar$, is a basic constant which is assigned a value of about 10^{-27} erg-sec. It follows that the wave function must always give some degree of uncertainty or ambiguity to either the position or the momentum of the parts of the system.

$$\phi(\bar{p}_1, \ldots, \bar{p}_n) = h^{-3n/2} \int e^{\frac{i}{\hbar} \sum\limits_{k=1}^{n} \bar{p}_k \cdot \bar{q}_k} \qquad (17)$$
$$\psi(\bar{q}_1, \ldots, \bar{q}_n) d\bar{q}_1 \ldots d\bar{q}_n$$

Figure 11. Transformation of the wave function from "configuration space" to "momentum space."[24]

This is commonly interpreted to mean that both P and Q have exactly defined values, but that the nature of experimental measurement is such that any measurement of P must involve some inevitable unpredictable disturbance in Q, and vice versa. Therefore, one can never obtain exact knowledge of both P and Q at once, even though they do possess exact values. The example is often given that if we use a high energy photon to determine the position of an electron, we can obtain an accurate value, but at the same time we must produce a large and uncertain change in the momentum of the electron. On the other hand, if we use a low energy photon we will not affect the momentum as greatly, but we will also not be able to obtain as accurate a value for the electron's position.

This interpretation of the uncertainty principle is widely held among scientists, particularly those outside the field of physics, such as biologists and chemists. However, it is actually not correct. If the theory is closely examined, we find that it is the *possibility* of making a measurement, rather than the measurement itself, which causes the uncertainty. According to Werner Heisenberg, "what is meant by the inevitable encroachment of the measuring process on the phenomenon is the possibility of measurement, i.e., the existence of the measuring device. For it is this possibility which . . . leads to the uncertainty relation."[25]

In order to clarify these rather obscure matters we shall not rely upon popular verbal interpretations of quantum theory. Rather, we shall work directly from the mathematical formalism which constitutes the theory itself, and see where it leads. The implications of the uncertainty relation will become clearer once we have analyzed what is known as the "reduction of the wave packet." For the moment we simply note that if the wave function is taken as the most complete description of the system, then the uncertainty relation implies that nature is *inherently ambiguous.*

What does it mean to say that a physical situation is inherently ambiguous? Since the constant, ℏ, is very small, it is possible for both ΔP and ΔQ to be smaller than the size of the experimental errors which one would inevitably expect to make in a practical measurement of momentum or position. One possible practical resolution of this question is therefore the proposal that this ambiguity is an aspect of nature with which we have no direct experience because its magnitude is so small as to be imperceptible. One might argue that only wave functions possessing such a practically unobservable degree of ambiguity should be admitted as valid descriptions of nature.

The Reduction of the Wave Packet

However, it is quite possible for small ambiguities of the order of magnitude of ℏ to magnify greatly with the passage of time if the wave function is assumed to vary according to the Schrödinger equation. This ambiguity may readily increase without limit, producing situations like that illustrated in Figure 21 or worse. For this reason, the theory must admit wave functions with very large ambiguities as valid descriptions of reality.

The basic question, "What is the nature of consciousness?" enters more directly into the theory of quantum mechanics than it did into the previous theories we have discussed. Those theories made no explicit reference to consciousness, and the question arose only when we considered the application of the theory to human life. In quantum mechanics, however, this question arises as soon as we try to understand how the theory describes matter, even though the study of life may not seem to be directly involved. This is due to the following two properties of quantum mechanical descriptions of nature:

(1) These descriptions are accepted as being complete. That is, they leave out no observable feature of the situation which is being described.

(2) Some of these descriptions display large ambiguities in the location and disposition of material objects.

These properties conflict with the following elementary property of conscious existence. This property is the most direct and immediate feature of our experience. We state it in some detail only because it may tend to be forgotten by a person who is bewildered by the formidable mathematical manipulations and abstract concepts of modern science.

27

(3) Each person has individual conscious awareness of thoughts, feelings, and sense perceptions. This awareness is definite for large scale objects. We may not be able to perceive what atoms are doing but we can see whether a table is on one side of a room or the other. Also, this awareness continues through time. Each person was consciously aware in the past, and his present memories, although imperfect, are consistent with his actual past conscious experience.

Since our experience is definite, (1) must be contradicted if we observe a situation corresponding to a wave function with large scale ambiguities. We see a definite situation and thus acquire information which was not present in the wave function. This becomes even worse if we put a human being into the picture and require quantum mechanics to describe him. If his wave function becomes ambiguous as in Figure 21, then (1) completely denies (3).

The standard quantum mechanical solution to this dilemma has been to invoke what is known as the reduction of the wave packet. In order to explain this we must describe two additional features of the quantum theory: the superposition principle and the normalization of the wave function. We will illustrate these features by considering a wave function, $\psi(q)$, for a mechanical system consisting of a single particle constrained to move along a straight line. The variable, q, will denote the position of the particle along the line.

First, let us define more clearly what is meant by saying that $|\psi(q)|^2$ represents the probability that the particle will be found in the position, q. This is illustrated in Figure 12. The probability that the particle will be found between the points a and b on the line is defined to be equal to the area under the curve, $|\psi(q)|^2$, between a and b.[26] This probability will therefore be equal to the difference between a and b multiplied by the average height of the curve between these two points. Likewise, the total area under the curve is equal to the probability that the particle will be found somewhere on the line. Since it must definitly be found somewhere, this probability is conventionally set equal to 1, meaning 100% certainty. It is therefore required that the total area under the curve, $|\psi(q)|^2$, must equal 1. This is called the condition of *normalization*.

If $\psi_1(q)$ and $\psi_2(q)$ are wave functions and u and v are numbers, then we can define a new function, $\psi(q) = u\psi_1(q) + v\psi_2(q)$, by ordinary multiplication and addition. As long as this new func-

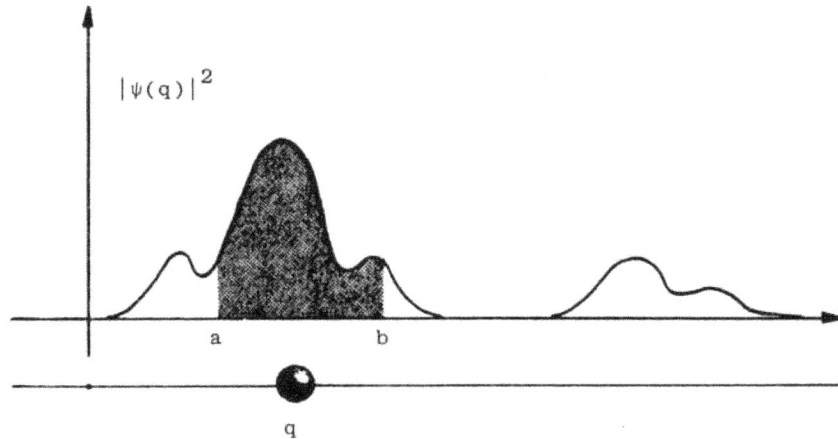

Figure 12. The probability that q lies between a and b.

tion is normalized, it is accepted as a valid wave function. (This will be true if $|u|^2 + |v|^2 = 1$.) This is called the *superposition principle*.

Figure 13 illustrates a particular example of this principle. In this figure, the wave function, ψ, is shown with two peaks, one having an area of ¾ and the other an area of ¼. Let us suppose that each peak is very narrow, but that the separation between them is large (as is shown in the figure). Then ψ may be interpreted as conveying the ambiguous information that the particle is near A with probability ¾ and near B with probability ¼. The wave functions ψ_1 and ψ_2, on the other hand, each give fairly unambiguous information about the location of the particle. (The wave function in Figure 10 can similarly be broken down into the superposition of two comparatively unambiguous wave functions concentrated near (a) and (b), respectively.)

This example gives some indication of how a wave function with a large ambiguity can be broken down into a superposition of wave functions, each of which separately has only a small ambiguity. The "reduction of the wave packet" consists of the following. A wave function that can be expressed as a superposition of unambiguous wave functions suddenly changes into one of them *at random*. That is, one of the functions entering into the superposition is randomly selected from among the others according to a certain assignment of probabilities, and put in place of the original function. In the example in Figure 13 this would entail randomly replacing ψ by either ψ_1 or ψ_2 with the respective

29

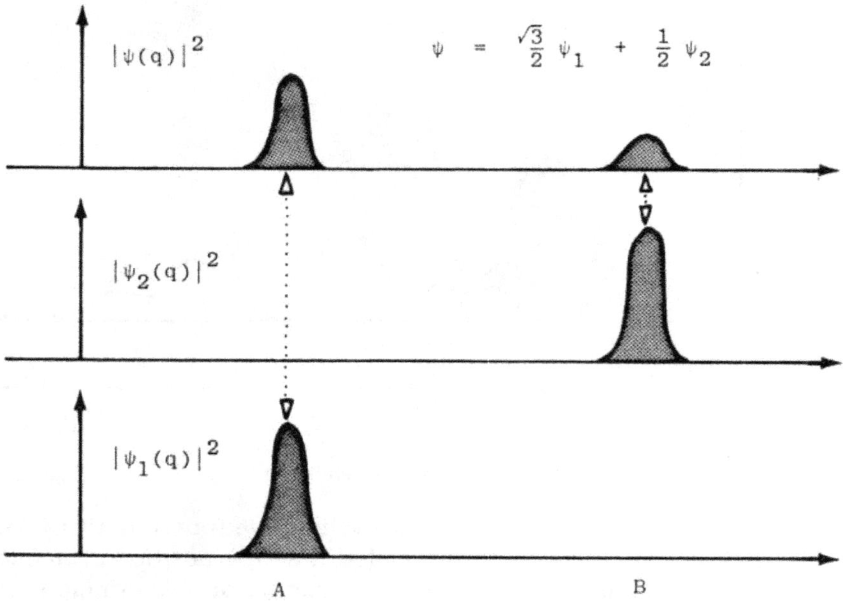

$$\psi = \frac{\sqrt{3}}{2}\,\psi_1 + \frac{1}{2}\,\psi_2$$

Figure 13. The superposition of two wave functions.

probabilities of ¾ and ¼. The standard solution offered by the theory of quantum mechanics to the dilemma of ambiguity is to suppose that *purely random* or *causeless* changes actually occur in nature corresponding to this sort of random replacement of the wave function.

As such, this process of random replacement or "reduction" of the wave function must be regarded as an additional quantum mechanical law of nature expressing how phenomena develop with the passage of time. It is outlined in more general mathematical form in Figure 14. Following the account of von Neumann, we shall also refer to this as process 1.[27] (The law of transformation expressed by the Schrödinger equation shall be referred to as process 2.)

This process can be invoked to solve the problem of ambiguity if we assume that it always occurs before any observer has an opportunity to inspect any system which has a wave function with large ambiguities. It solves the problem by the simple device of replacing each undesirable ambiguous state of affairs by an unambiguous one before anyone has a chance to witness it. However, this is not a very satisfactory solution.

Suppose that A represents an observable quantity with a spectrum, a_k (k = 1,2,3,...), of possible observed values. Let P_k denote the operation of projecting a function onto the space of functions for which A takes on the exact value, a_k.

Process 1 is defined to be the *random* substitution,

$$\Psi \rightarrow \frac{P_k \Psi}{\|P_k \Psi\|} \text{ with probability } \|P_k \Psi\|^2.$$

(For example, if A is a [Q/a], where Q is a coordinate of a large object, then the values, $a_k = ka$, represent positions of the object to within an accuracy of $\pm\, a/2$.)

Figure 14. Formal definition of the reduction of the wave function. [28]

First of all, it throws out the traditional scientific conception that the phenomena of nature occur according to definite causes, and introduces the idea that "pure chance" is an inherent aspect of reality. In the previous theories of physics, "chance" had entered only as a measure of the ignorance of a human observer about the details of observed phenomena. It was always assumed that the phenomena themselves occurred in a perfectly definite way according to precise laws of cause and effect. Perhaps the most striking example of this is the classical theory of statistical mechanics, in which probabilities were used to describe the positions and momenta of the atoms in a gas or liquid. Probabilities were introduced because it was impossible for anyone to acquire knowledge of the exact situation of each one of the practically innumerable atoms involved. However, it was nonetheless supposed that these atoms did have definite situations and were behaving in accordance with the exact laws of motion that were accepted as valid at that time.

Process 1, on the contrary, is assumed to operate without any cause whatsoever. As we shall point out in Section V, no one has ever been able to provide any clear understanding of just what "action without cause" really means. All attempts to characterize pure chance mathematically have failed, either because of circular reasoning, or because they did not confront this problem directly.[29] Actually, if a phenomenon occurs without any cause, then that phenomenon can be regarded as a feature of the absolute truth, for the absolute truth is defined to be the cause of all causes.

31

Viewed in this way, the initial conditions, the Schrödinger equation, and all the outcomes of occurrences of process 1, could be regarded as the complete absolute truth according to quantum theory. However, this is even more awkward than the picture of the absolute truth as initial conditions plus laws of motion which we discussed in the last section.

In addition, there is no satisfactory way of understanding why process 1 should occur at any one particular time and not another. As described by the physicist Eugene Wigner, "the 'reduction of the wave packet' enters quantum mechanical theory as a *deus ex machina*, without any relation to the other laws of this theory."[30] It gives every indication of being simply an arbitrary, ad hoc adjustment which is introduced solely to save the theory from the contradiction between points (1), (2), and (3).

We propose that a simpler and more realistic solution to the problem of ambiguity is to suppose that quantum theory is incomplete. That is, we propose that point (1) should be rejected. According to this understanding, the wave functions and the Schrödinger equation may be regarded as providing an approximate description of certain natural phenomena under limited circumstances. Because they fail to take into account many important features of reality, they cannot provide a complete description of natural phenomena in general. Thus, they may be expected to yield descriptions of phenomena which become increasingly inaccurate or ambiguous with the passage of time.

In the remainder of this section we shall discuss several practical examples illustrating the different aspects of the theory which we have described thus far. In particular, these examples should bring out very clearly how quantum theory gives rise to ambiguous descriptions of nature conflicting with the fundamental features of sentient existence outlined in point (3). It should also become apparent that process 1 is thoroughly untenable as a solution to this problem.

Practical Illustration of the Theory
The Phenomenon of Alpha Radiation

A familiar experiment discussed in elementary science books is the observation of the emission of alpha particles from radioactive atoms placed in a Wilson cloud chamber. In this experiment, a small radioactive object is placed in a glass chamber. When the humidity within the chamber is adjusted properly, thin lines of fog are seen to extend out from the object into the surrounding air. The customary explanation of this is that certain atoms in the object are unstable. Such an atom can suddenly expel a small particle, called an alpha particle, with a very high velocity. As this particle moves through the air of the chamber, it collides with atoms along its path and ionizes them. Since the humidity is nearly at the saturation point, droplets of water condense about these ions, leaving a visible trail of fog which marks the path of the particle.

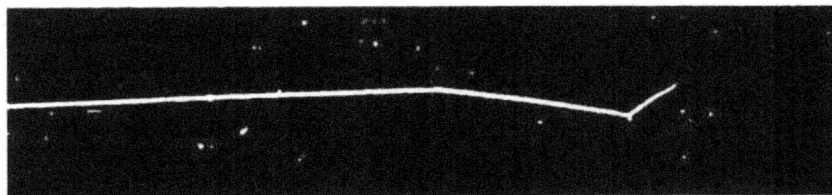

Figure 15. Photograph of an alpha particle trail.[31] The particle is thought to have been emitted from an atom of radon gas at the left hand end of the trail.

This explanation is based roughly on the ideas of classical 19th century physics, and it is now rejected. Therefore, let us consider the quantum mechanical explanation. In order to do this we must first describe the appropriate quantum mechanical wave function. As this example is intended to illustrate the basic features of the quantum theory, we will make some standard idealizations of the physical situation for the sake of simplicity.

First, let us consider the alpha particle by itself. The radioactive atomic nucleus is normally represented as a small spherical barrier which tends to hold the alpha particle inside. The theory of quantum mechanics requires that this particle should be described by a wave function, and the solution to the Schrödinger equation in this situation shows that this wave function tends to "tunnel" through the barrier in all directions, producing a three dimensional spherical wave. This can be compared formally with the situation of a light wave trapped within a hollow shell of partially reflecting glass: as the light bounces back and forth within the shell some of it is transmitted and escapes, while some of it is reflected and continues bouncing.

33

The spherical wave is depicted in Figure 16. This wave function is easy to represent graphically because it is three dimensional. However, it is not very meaningful because it does not take into account the existence of any other matter with which the alpha "particle" might interact. If we add coordinates describing additional matter, such as the air in the cloud chamber, then the wave function for the system will have to be a multi-dimensional mathematical object. As we have pointed out, three dimensions must be added to the domain of the wave function for each additional particle considered in the physical system.

Let us enlarge the system by adding some additional atoms in the vicinity of the radioactive atom. According to quantum theory, these atoms can exist at various energy levels, ranging from the lowest or "ground state," to higher "excited" states. We shall assume that the atoms are initially in the ground state. The classical conception of alpha radiation is that at some subsequent time a

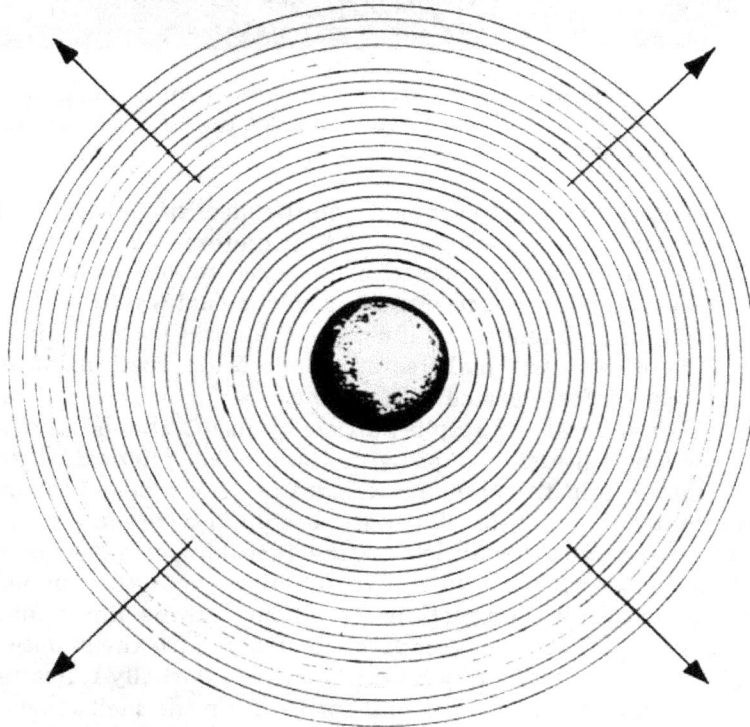

Figure 16. The wave function of an alpha particle.

34

particle will fly out of the radioactive atom and hit some of these atoms, raising them to excited energy states. This is shown in Figure 17, where four such "target atoms" have been added. One of them has been hit by the particle and raised to an excited state.

If we assume that these atoms are so far apart from one another that they do not interact, then it is possible to visualize the wave function corresponding to the quantum mechanical description of this system.[32] By a mathematical transformation, the multivariable wave function for the system of alpha particle plus target atoms can be expressed as a collection of alpha particle wave functions.[33] Each of these functions corresponds to a particular one of the many possible arrangements of energy eigenstates in which the target atoms could be situated.

In Figure 18 we have tried to take advantage of this mathematical transformation as a means of representing the wave function pictorially. For simplicity, let us consider the target atoms to be either "excited" or "unexcited." In the figure, five of these alpha particle wave functions are depicted graphically as waves on five planes stacked one on top of the other. (Each plane actually represents three dimensional space.) The bottom plane shows all four target atoms unexcited. In this plane the alpha wave is similar to the one in Figure 16. In each of the remaining planes, one atom is excited ($+$) and the others are unexcited ($-$). In these planes the alpha wave extends in a narrow beam from the excited atom, as shown in the figure. Taken together, the wave functions in these "planes" constitute the complete wave function of the system at one moment in time.

The total wave function for the system is thus equivalent to the complete collection of these alpha waves for all possible combinations of excited and unexcited states of the target atoms. Since the wave function is an abstract mathematical construct, and

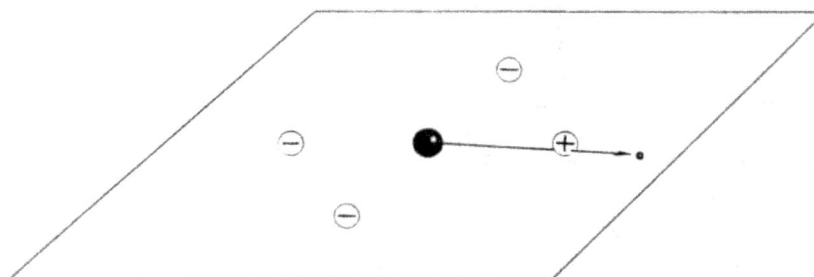

Figure 17. The "classical" picture of alpha radiation.

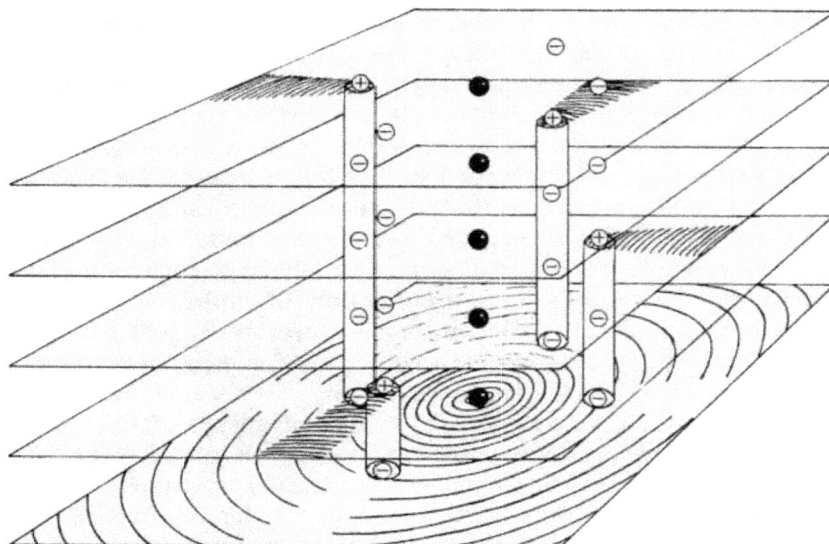

Figure 18. The corresponding quantum mechanical picture.

in no way "real," this is as good a way as any of representing it. This representation has the value that it enables us to easily visualize how the wave function changes with time according to the Schrödinger equation. We should emphasize that the planes do not represent different alternative situations which may have taken place. Rather, we can interpret Figure 18, considered as a whole, as a complete description of the physical situation of the system at one time, according to the quantum theory.

The action of Schrödinger's equation in this situation can be understood as follows. Suppose that initially all of the target atoms are unexcited. This means that the initial wave function should have the spherical wave of Figure 16 in the bottom plane and nothing at all in the planes representing excited atoms. Then, as this wave moves past each unexcited atom, part of it flows out from the site of that atom in the particular upper plane where it is designated as excited. Apart from this transfer from one plane to another, the alpha waves can be thought of as moving like ordinary ripples on water. Figure 19 illustrates the transfer of the alpha wave from one plane to the next.

This description can be formally deduced from the Schrödinger equation for the system. If we had many target atoms distributed around the central radioactive atom, then the picture would be similar but more complicated. In this case there would be many planes[34] corresponding to different arrangements of excited and unexcited atoms, and the alpha waves could be visualized as flowing similarly from plane to plane. This is shown in Figure 20 for two of the many possible planes in this representation.

We have described this wave function in some detail in order to raise the question, "Just what does this have to do with reality?" First, let us consider the probability interpretation of the wave function. In this representation, the absolute value squared of the alpha wave at a certain point in one of the planes is interpreted as measuring the probability that the alpha particle is located at that point and that the atoms are in the pattern of energy states designated by that plane. The total probability for the alpha wave in the plane is therefore the probability that the target atoms will be in that particular pattern of energy states.

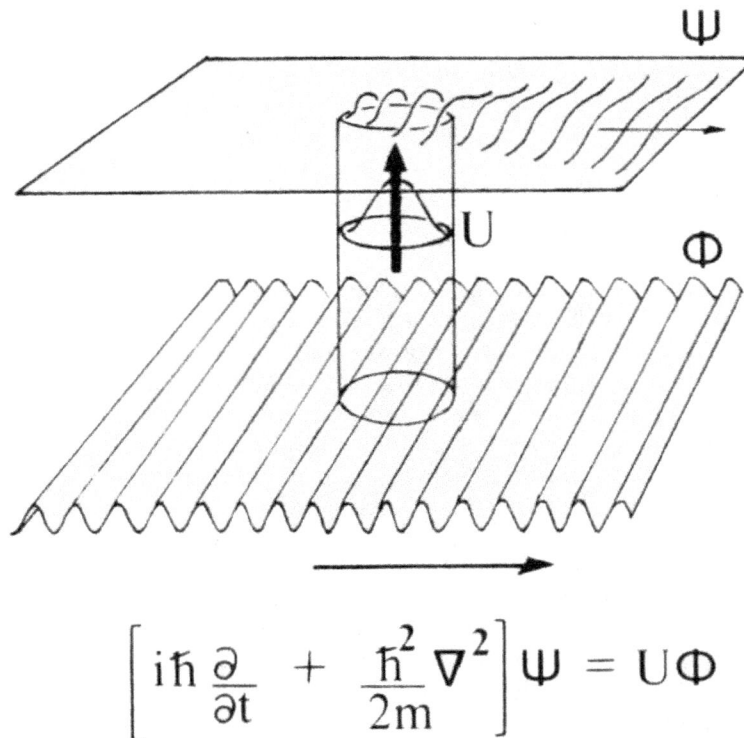

$$\left[i\hbar \frac{\partial}{\partial t} + \frac{\hbar^2}{2m} \nabla^2 \right] \Psi = U\Phi$$

Figure 19. Pictorial representation of the solution to the quantum mechanical equation showing the interaction of a moving "particle" with an atom.

37

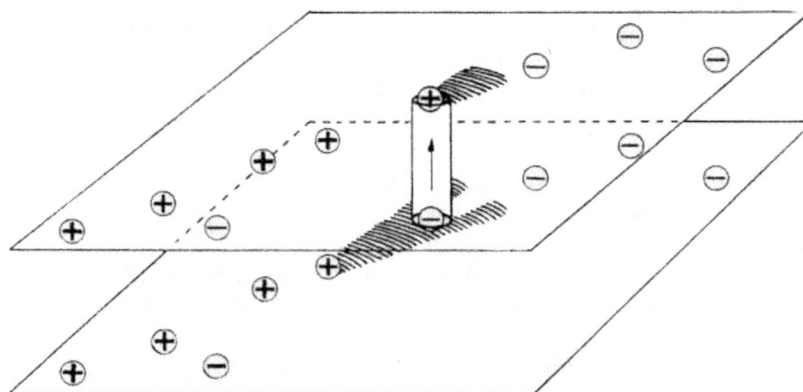

Figure 20. Part of the total wave function for a system with many target atoms.

We can see from Figure 18 that in the case where there are four target atoms, the wave function gives equal probability to the possibility of each atom being excited by the alpha particle. If there are many target atoms, then each plane designating a string of excited atoms extending in a nearly straight line from the radioactive atom will have nearly equal probability in the wave function.[35]

Now, if the wave function is taken as a *complete description* of the condition of the system at a given moment, then how is this to be understood? Should we suppose that in nature the atoms of gas in the cloud chamber will be ionized in the form of a trail, but that nature is unclear as to which one out of many trails pointing in different directions will be actually present? That is the implication of the wave function, since each possible ionized trail is represented there with nearly equal probability.

Another feature of the wave function is that it does not indicate any particular time at which the atom "decays" and produces a trail. Certainly our mental conception of this phenomenon is that at some one moment in time a trail of ions begins to extend out from the radioactive atom in one definite direction, and that a thin line of fog then condenses about these ions. Such a definite trail is shown in the photograph in Figure 15, and it was presumably formed at a definite time.

However, the wave function neither indicates a definite time nor a definite direction for the trail. Starting at the initial time when all of the target atoms are assumed to be unexcited, the alpha waves immediately begin to flow through the different planes which represent different possible ion trails. Within a brief fraction of a second, there will be alpha waves in each plane representing a trail, and this implies that each trail will have some positive probability. All that then happens with the passage of time is that the probabilities of the trails all increase as the alpha waves build up in each plane. In fact, the very idea of definite events happening in time is foreign to this quantum mechanical model. Here there is simply the flow of "probability waves" past varieties of static alternative situations—in this case, possible ion trails of various lengths extending in various directions. There is no time at which the atom decays.

The Introduction of a Human Observer

Our discussion of alpha radiation is still incomplete, however, because we have not formally taken into account the observation of the ion trails by a conscious person. In order to take this into account quantum mechanically, we would have to introduce coordinates for the atoms composing the body of the person and his experimental apparatus. We would have to assume that everything about the observer could indeed be described by a wave function—a complex valued mathematical function of those coordinates.

In the previous section we pointed out that the theories of the 19th century, which described the world as a collection of simple objects such as vectors or atomic particles, could not be said to account for consciousness. In quantum theory, however, it is not at all clear just what the world is supposed to consist of. It cannot be made of definite particles, since the Heisenberg uncertainty principle implies that there must be ambiguity in position or momentum in even the most complete quantum mechanical description of matter. On the other hand, it cannot correspond in any one-to-one way with the wave function, since this is a very artificial mathematical object. This lack of clarity hardly makes quantum theory more qualified as a theoretical explanation of consciousness than were the nineteenth century theories.

Nonetheless, the common assumption is that if a theoretical scheme can correctly describe the behavior of matter, then it can also account for the conscious experience of persons whose bodies are so described. Therefore, we would like to see whether or not quantum theory by itself can correctly account for the phenomena of nature which are involved with living beings. As the theory stands, this requires that these phenomena must be completely accounted for by wave functions which change in time according to the following factors:

(1) The "reduction of the wave packet." (Process 1.)
(2) The Schrödinger equation. (Process 2.)
(3) Initial conditions. (Boundary conditions may also be involved if the behavior of matter outside of the system is significant.)

If this is not possible, then other principles must be involved in the phenomena of nature. (There are some discussions of this subject which make misleading use of a mathematical construct known as a "mixture" or "density matrix." We will discuss this when we describe the views of Daneri, Loinger, and Prosperi in the Appendix.)

So, let us suppose that we now add to the system all the apparatus of the cloud chamber, plus the body of a human observer. This means that the Schrödinger equation will have to be augmented by terms describing the interaction of the various coordinates describing the apparatus and observer. Also, terms will have to be added describing the interaction of the target atoms with one another. In the analysis given thus far, it has been assumed for simplicity that these atoms don't interact among themselves. However, it is by the mutual attraction of the water molecules about the ionized atoms that the visible fog droplets marking the trail are supposedly formed.

If all of these things are added, then the wave function becomes even more multi-dimensional than before, and the Schrödinger equation becomes much harder to solve. Actually, we cannot even begin to solve Schrödinger's equation for a human body, and so we cannot actually say what such a solution would yield. In fact, the process of condensation of saturated water vapor to form liquid droplets is also beyond the reach of present mathematical techniques. Condensation has only been studied for highly idealized and simplified models, such as the Ising model, and even there the analysis is very difficult.

For this reason, it is necessary to make some assumptions. Since we are trying to show that quantum mechanics is incomplete, let us suppose that Schrödinger's equation does correctly describe the behavior of a human body as we see it. (This drastic assumption is quite doubtful, as we shall see later.) If this is wrong then quantum mechanics must be incomplete, which is our proposed conclusion, anyway. Let us suppose, in fact, that the following sequence of events will be correctly described by the Schrödinger equation: start with a wave function for a human observer watching a cloud chamber which contains a specific trail of ions. Suppose that the trail has gotten there somehow, but that it has not yet had an opportunity to interact with anything else. Then, as time passes, a trail of droplets will condense about the ions, light will reflect from the trail, and the human observer will see it and react in some fashion.

If we make these assumptions, then we can modify the analysis of the system of alpha particle plus target atoms to take into account the observer. This works out roughly as follows. The interaction of the alpha particles with the target atoms will be nearly the same as before, since the mutual interaction of the target atoms and the subsequent reflection of light will not have much effect on this. (The interaction of the target atoms with one another will occur according to much weaker forces than their interaction with the alpha particle.) Therefore, we will still obtain a wave function giving nearly equal probability to many different trails.

Now, however, the sequence of events from trail to observer implies that the wave function will represent many different perceptions within the mind of the observer. As time passes, many different positions and mental states of the observer will come to be equally represented by the wave function.[36] Thus, the wave function will depict the kind of ambiguous situation shown in Figure 21.[37] Here we should again emphasize that the separate "planes" do not represent alternative situations—all of the planes, taken together, represent the complete situation of the "observer" at one time.

In the figure, the part of the wave function in each plane is shown depicting the observer in essentially one position. However, we might expect many different positions to be represented. If the observer had planned to go out for lunch after seeing a track, then some of the planes would show him still waiting for the track to occur, while others would show him at various stages of eating lunch. These are all aspects of the one multi-dimensional description of nature provided by the wave function.

If the wave function is taken as a complete description, then this situation contradicts the condition of consciousness mentioned previously. As we have pointed out, the standard solution

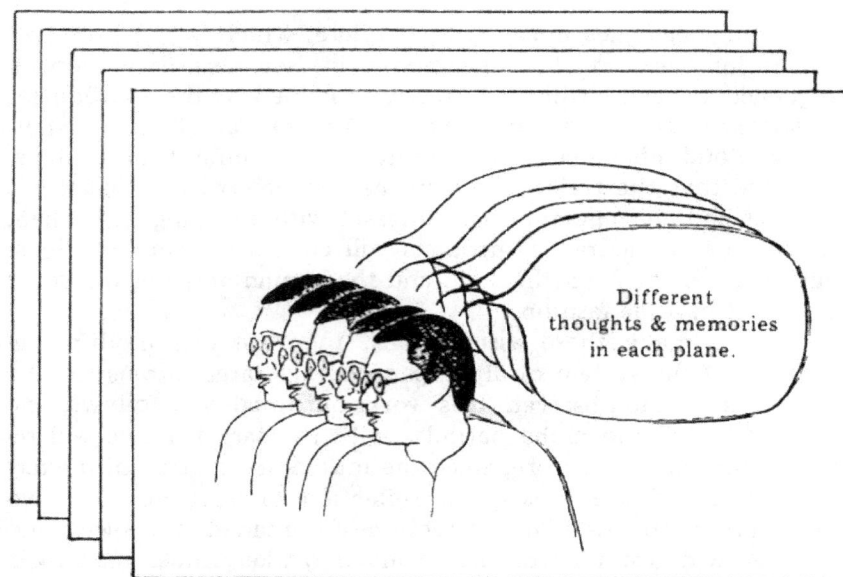

Figure 21. The wave function for the human observer. Here each "plane" represents a particular situation of the observer. The total wave function gives nearly equal weight to many such planes.

to this dilemma is to invoke the reduction of the wave packet and suppose that nature abruptly changes by pure chance from an inherently ambiguous situation to an unambiguous one. However, if we invoke this process once, say at time t, then we have the problem that the wave function for the observer was ambiguous before t, and this still contradicts his consciousness. This is actually equivalent to saying that the observer has no definite past before the time, t: he is "created" at this time by pure chance along with his memories of a nonexistent past!

In order to avoid this problem, it would be necessary to bring in process 1 repeatedly at short enough time intervals to prevent the wave function from ever developing ambiguities in the condition of the observer. This is because the steady flow of the wave function from the "unexcited" plane to the planes representing tracks produces a steady creation and amplification of ambiguity within the system at all times. Figure 22 illustrates the view of nature which is involved here. Each S represents a period of duration, Δt_k, in which the Schrödinger equation governs the wave function. The R_k's represent interruptions by process 1. The timing of these interruptions is presumably "causeless."

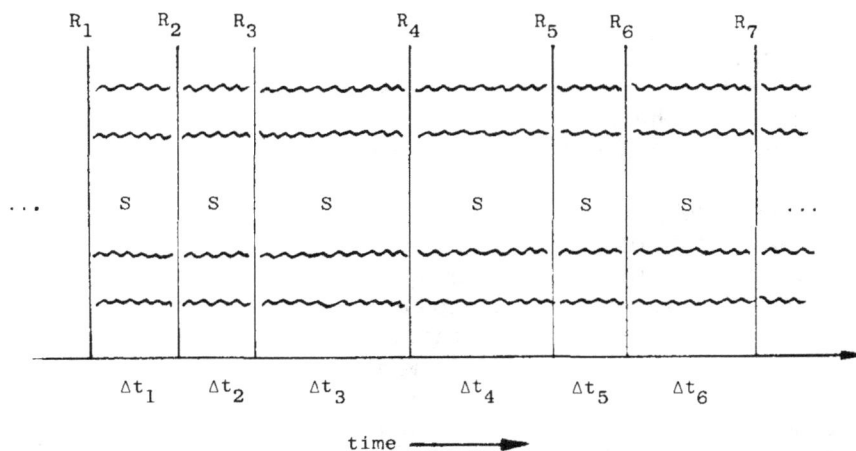

Figure 22. The development of the wave function in time. The S's refer to intervals in which the Schrödinger equation operates. The R's represent interventions of process 1.

Here the Δt's cannot be large enough to allow any ambiguity to develop in what the observer "perceives." On the other hand, if the Δt's are too small, then the development of probabilities with time predicted by the theory is drastically altered. When the Δt's become very small, the process of radioactive decay which we are considering is slowed down, and it stops completely if they are reduced to zero.[38]

Within these two limits, the Δt's can be arbitrarily chosen. Are we to suppose that some random process in nature is actually producing alterations of this kind, and at these intervals, in the actual natural situation? If so, then what is this process? Why are the Δt's chosen as they are? If consciousness is merely an evolved by-product of nature, as modern scientific doctrine has it, then the Δt's should be the same throughout nature whether consciousness happens to be present or not. In particular, they should have been the same during the billions of years which are supposed to have passed before consciousness evolved.

In that case, we are confronted with a process which repeats itself due to no cause whatsoever at a sufficient rate to prevent the Schrödinger equation from rendering consciousness contradictory, and thus allow for the possibility that it might "evolve" from an unconscious state! We do not know at what intervals this process repeats itself. It involves absolutely random changes which take place abruptly over large distances in (multi-dimensional!) space

43

and are thus by no means atomic in scale. If consciousness is just a material byproduct, then this process must be independent of consciousness. Yet, we are forced to postulate it simply in order to accommodate the principle that the wave function is a complete description of nature to the perceived fact that consciousness exists.

It would certainly seem that in postulating such a process we are deviating both from common sense and all traditional procedures of scientific reasoning simply in order to retain the satisfying viewpoint that we have a complete description. Perhaps it would be more reasonable to admit that the wave function is not a complete description—that there are very important aspects of reality which it does not touch upon. Nonetheless, it is precisely this process which lies behind the celebrated scientific conclusion, taught to all students in both high school and college, that the laws governing nature are based on "pure chance."

Since consciousness plays such an important role here, let us again consider its basic nature. It is sometimes claimed that consciousness is something "subjective," whereas science is solely concerned with "objective facts." It is asserted that an entity can only be regarded as real if several people can all perceive that entity and agree upon what they have seen. Since the consciousness of one person is only experienced by that person alone, it is argued that the existence of consciousness is not a valid topic for rational scientific discussion. Rather, since the gross movements and other measureable features of the body can be seen and agreed upon by many simultaneous observers, it is said that these form the sum and substance of the existence of a living being.

However, this argument is not valid. First of all, since it deals with "perception" and "agreement," this argument presupposes the existence of the very thing which it seeks to reject. Second, the fact that a thing cannot be seen does not imply that it does not exist. It is a common practice in science to infer the existence of certain entities from the presence of observable phenomena which are associated with them. For example, no one has ever seen atoms, but their presence is inferred from certain phenomena which can be measured and agreed upon.[39] Likewise, a given person can directly perceive his own conscious existence and note that it is intimately tied up with an intricate constellation of thoughts, feelings, perceptions and desires. He can see that these things are associated with certain actions of his body which can be observed by others. If he then observes the body of another person and perceives similar patterns of activity, it is not unreasonable for him to infer that those actions are likewise associated with the thoughts and feelings of another conscious entity like himself.

It is thus perfectly in accord with the practices of scientific reasoning to suppose not only that consciousness exists, but that there are many beings possessing conscious self-awareness. The existence of consciousness is thus just as much an objective fact as is any other "fact" considered in scientific discussions. The philosophy of solipsism, which attributes consciousness to only one person, certainly violates common sense (if nothing else), and the proposal that there is no conscious awareness at all is plainly untrue.

Evidently something exists which is giving rise to consciousness, and that something lies beyond the grasp of science as it is presently constituted. We have already observed that the existence of consciousness could not be accounted for by past physical theories. Here, however, the basic theory not only fails to account for consciousness, but the very existence of consciousness cannot be accommodated by the theory without the introduction of absurd stop-gap measures. A different description of nature is clearly needed.

We should note that the phenomena we have been discussing are not limited to the particular case of alpha radiation. Rather, they are typical of the quantum mechanical situation in general. The domain of the wave function is the multi-dimensional space of coordinate configurations, and so it is natural for this function to spread further and further through this space under the action of the Schrödinger equation. It is therefore not at all surprising for the function to spread to parts of the space representing macroscopically different configurations, and thus produce a situation corresponding to a large scale ambiguity.

Whenever events on the molecular level produce macroscopic effects we should expect the wave function to develop such ambiguities. This is due to the "amplification" of the inherent small scale ambiguities of the wave function which are indicated by the Heisenberg uncertainty relation. In particular, we might expect this to happen in the bodies of living organisms if these are to be regarded as automatons governed by the detailed processing and storage of information at the molecular level. It is, of course, much more difficult to analyze the wave function in enormously complex cases such as this than it is in the relatively simple situation which we have discussed here.

45

The Paradox of Einstein, Podolsky, and Rosen

Let us return for a moment to the process of repeated reductions of the wave function and consider the size of the phenomena affected by them. In our example, the ion trails will not affect the gross ambiguity of the wave function until perceptible droplets of fog have condensed about them. It follows that the wave function will become ambiguous with respect to the distribution of these trails over several inches of space before the action of process 1 causes a particular trail to be selected. From this we can see that the world cannot at all be thought of as being composed of distinct "particles" in the quantum mechanical view. We must suppose that at one moment in time the "alpha particle" is somehow spread over space in a way that can only be described by a multidimensional scheme such as the one depicted in Figure 18.

A striking example of how far an ambiguity in a wave function can extend in space is given by the famous paradox of Einstein, Podolsky, and Rosen.[40] We shall briefly summarize a description of this paradox given by the physicist David Bohm in terms of the quantum mechanical property called "spin."[41] In quantum mechanics it is supposed that an electron can be thought of as possessing a magnetic field which can be aligned in a particular direction. (This can be compared to the "dipole" magnetic field of a bar magnet.) In this paradox, a situation is described in which two electrons fly apart. Their wave function unambiguously determines that the magnetic field of the first electron points in exactly the opposite direction of the magnetic field of the second electron. However, the situation is deliberately arranged so that the wave function is completely ambiguous as to what this direction is.

In principle, it is possible for the electrons to fly apart for many miles and then be detected by widely separated observers with instruments designed to measure how their paths are deflected by large magnets. At this point, their ambiguity in spin direction (as it is called) results in a large scale ambiguity in the perceptions of the observer, and thus process 1 must be invoked.

Prior to this reduction of the wave function, the two electrons cannot be thought of as separate "particles" in separate locations. In fact, they can only be thought of in terms of one multidimensional entity extending over miles of space. Yet, when process 1 takes place, this entire widely-extended wave function must be supposed to change abruptly and randomly into a new

46

wave function in which the spin direction at each location is
unambiguous.

WAVE FUNCTION BEFORE REDUCTION

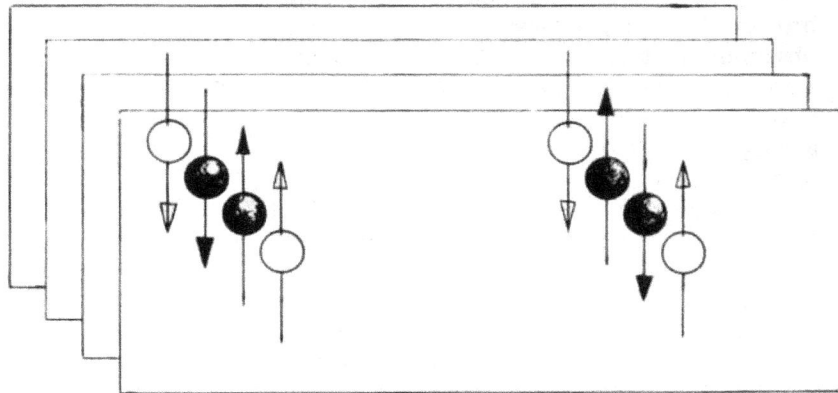

SUBSEQUENT WAVE FUNCTION (PROBABILITY OF 1/2)

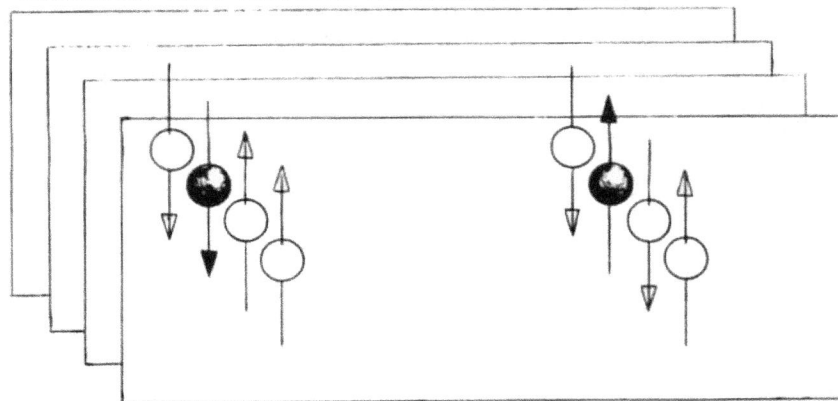

|←———— SEVERAL MILES ————→|

Figure 23. Sudden long range random change in quantum theory. This
figure represents the reduction of the wave function for two
widely separated electrons. Here, the different planes correspond
to different arrangements of electron "spin" in the z direction.

47

It is not possible to suppose that the spin directions could have been rendered unambiguous by a reduction of the wave packet before the electrons became widely separated. The reason for this is that the manner in which the wave function of the electrons is broken down by process 1 depends on the orientation of the observers' instruments.[42] As a result, if the electron wave function were to be reduced before the electrons became widely separated, it would have to be done in such a way as to anticipate the later arrangement of the instruments. This is a bit too much to expect of a causeless process.

Considerations of this kind led Albert Einstein to conclude that the quantum theory was unrealistic. In his paper with Podolsky and Rosen, he specifically proposed that quantum mechanics gives an incomplete description of reality. He suggested that the situation could be remedied by returning to a theory of the classical 19th century variety, in which there is an exact one to one relation between mathematical quantities and elemental constituents of matter ("elements of physical reality.")

However, he was never able to produce such a theory. The present scientific world view remains that of quantum mechanics— a world view in which ambiguities in nature may extend over miles of space, only to be suddenly, causelessly, and randomly removed. They are removed over large volumes of space instantaneously by an inherent process of nature, the only requirement being that this removal must take place before the ambiguities extend to a functioning human brain. Yet such a brain is simply another arrangement of matter—an arrangement which modern scientific opinion holds to have arisen in nature only recently.

IV.

Interpretations of the Quantum Theory

The Heisenberg-Bohr tranquilizing philosophy—or religion?—is so delicately contrived that, for the time being, it provides a gentle pillow for the true believer from which he cannot very easily be aroused. So let him lie there.

—Albert Einstein

In summary we can see that the complete world picture, as presented by the theory of quantum mechanics, consists of the following basic features:

(1) A very unclear and artificial description of the world as a multi-dimensional mathematical abstraction described in terms of nonexistent "particles."

(2) An equation consisting of a few terms which determines how this description changes with time.

(3) An arbitrary and not clearly defined ad hoc adjustment of this description which must be made repeatedly in order to free it from contradictions arising from (2).

In particular, (3) involves the assumption that abrupt, large scale changes occur in nature which have no connection with the other laws of the theory, and are not clearly defined as to either their timing or their exact form. These are attributed to "pure chance."

It is not surprising that many scientists have been disturbed by this world picture, and have sought ways of rectifying it. Some, such as Albert Einstein, have rejected the quantum theory altogether and called for a completely new theoretical system. Others have sought to retain the formal mathematical structure of the theory as it stands and find some interpretation which renders it more palatable.

In this section we shall discuss two interpretations of this kind. The first of these, called the Copenhagen or "conventional" interpretation, has been the most influential view of quantum mechanics. It is widely held among physicists. However, many are not very deeply concerned with its implications and accept it, according to one author, "just as most Americans would claim to believe in the Bill of Rights, whether they had ever read it or not."[43]

49

The second interpretation was presented by John von Neumann in his *Mathematical Foundations of Quantum Mechanics.* This view is significant in that it gives direct recognition to the role of consciousness in the quantum theory. All the other approaches we shall touch upon have evaded this issue in one way or another.

Just to illustrate the extent of the controversy and confusion the quantum theory has evoked, we will also briefly mention a number of other interpretations. Two of these are of a more technical nature and are discussed in the appendices. Of particular importance is the theory of Daneri, Loinger, and Prosperi. This theory is an unsuccessful but superficially convincing attempt to eliminate the problem of the reduction of the wave packet by exploiting the mathematical concept of a "mixture" of quantum mechanical states.

The Doctrine of Complementarity

Thus far in our discussion of quantum mechanics we have been adhering formally to the Copenhagen interpretation of the theory. In this interpretation, it is assumed that the quantum mechanical formalism which we have described provides a complete account of nature. The interpretation provides a philosophy, known as the principle of complementarity, which is intended to render this assumption intelligible. It was devised primarily by Niels Bohr and Werner Heisenberg during the twenties and thirties of this century.

Since the Copenhagen interpretation is based on the assumption that the quantum theory is universal and complete, it follows that the features of this view must correspond directly with the formal mathematical manipulations entailed by the theory. Here are some of these features:

(1) Nothing exists in nature which is not described by the wave function.
(2) Matter cannot be conceived of as composed of particles, waves, or anything else that can be visualized in three dimensional space.
(3) The world cannot be thought of as composed of distinct parts, as in the 19th century view. It must be viewed as a unit.
(4) This unit will exhibit different features depending on how it is measured. These features may seem to be mutually

contradictory, but there is no contradiction because the measuring apparatus is itself part of the unit being measured.

(5) For example, one measurement of an electron may seem to indicate a wave, and another may seem to indicate a particle. There is no contradiction because the apparatus of measurement is different in the two cases. One is not observing an "electron," but a total system of "electron," apparatus, and whatever else may be involved.

(6) When a measurement is made, an abrupt, unpredictable change is made in the measured system. This is called a "quantum jump," and it is said to occur by absolute chance.

(7) It is taken for granted that the macroscopic world of our everyday experience behaves in accordance with classical nineteenth century physics. All experimental measurements are presumed to end with the observation of a macroscopic apparatus obeying the classical nineteenth century physical laws.

Of these features, numbers (2) through (5) follow as logical consequences of number (1).[44] We can gain some appreciation of how this is so from an examination of figures 15 through 21 of the last section.

Number (6) refers to the "reduction of the wave packet," or process 1. We have noted that this process is sometimes attributed to the disturbance an observer must make in an observed system due to unavoidable clumsiness in his measuring technique. The example is often given that if one measures the position of an electron by bouncing a photon against it, the momentum of the electron must necessarily be disturbed in an unpredictable way by the collision. It should be clear from the discussion in the last section, however, that this is not at all the correct interpretation of the "quantum jump." The theory requires that these jumps must regularly occur even if the observer does nothing at all.[45] They must, in fact, spontaneously affect the total system of observer plus apparatus without the assistance of any outside interference.

The Copenhagen interpretation provides no explanation of these mysterious jumps. They are sometimes referred to in vague terms as the transition from "potential" to "actual." The essence of the Copenhagen view, however, is that these jumps must simply be regarded as inexplicable, elemental, and not understandable in terms of anything else. Thus, they form a part of the absolute truth as it is understood in this world picture.

In the quantum theory one normally only discusses the results of some experiment made on inanimate matter by means of some particular apparatus. As such, number (7) provides a practical basis for such discussions. If number (7) is regarded as being true in principle, however, then the implication is that quantum mechanics is an incomplete description of matter.[46] If quantum mechanics is to be viewed as complete, it must be possible to include both the apparatus and the observers in the quantum mechanical system. We have seen the unsatisfactory consequences this entails.

The principle of complementarity enunciated by Niels Bohr is essentially a restatement of points (2) through (7) in philosophical language. According to Bohr, the world possesses pairs of complementary features which may seem to contradict one another but are actually not contradictory because they cannot be observed at one and the same time. Perhaps the most famous example of a pair of complementary traits is the so called wave-particle duality, in which matter is attributed both wave-like and particle-like properties, but cannot be seen to exhibit both of them simultaneously. Position and momentum, and spin in the x and y directions are other examples of complementary pairs.[47] In the mathematical system of quantum mechanics, these pairs of traits arise because of the different ways in which the wave function can be broken down by process 1. Bohr's philosophical stance is that the pairs of complementary features are to be regarded as absolute aspects of reality that must simply be accepted as they are, and not analyzed further in terms of any other concepts.

As it is normally stated, the subject matter of the Copenhagen interpretation involves an observer who is studying some atomic phenomenon by means of some macroscopic apparatus. It describes how different arrangements of the apparatus will enable the observer to perceive different complementary traits of the phenomena. Taken literally, it is simply a description of how certain actions will lead to certain results when carried out by a conscious observer who has freedom to act. This would seem to imply that the observer himself must be assumed as given, and cannot be further analyzed by the theory. This concept is, in fact, strongly suggested by W. Heisenberg's statement that "when we speak of a picture of nature provided by contemporary exact science, we do not actually mean any longer a picture of nature, but rather a picture of our relation to nature."[48]

52

However, the assumption of universality requires that the observer himself must be completely described by a wave function. If this wave function must be interpreted in turn as a description of what another observer will perceive when he interacts with the first observer, then this leads us to an absurd situation. We either have two observers, each of whom is completely described to the other by a wave function, or we are faced with an infinite regress of observers. The problem here is that the theory requires the existence of a conscious observer and is required at the same time to explain him in physical terms. Yet, the theory is completely unable to explain consciousness. Unfortunately, Bohr did not confront this question squarely. Bohr's characteristic philosophical stance has been to declare that any question which cannot be answered by the quantum theory is in principle unanswerable and should not be asked. He has thus said that "an analysis of the very concept of explanation would, naturally, begin and end with a renunciation as to explaining our own conscious activity."[49]

Bohr regarded the principle of complementarity as a universal explanatory device, and he applied it to many things that are beyond the immediate realm of physics. In particular, he proposed that the phenomena of life and the molecular interactions occurring in living organisms form a complementary pair.[50] These two aspects of the organism cannot both be observed at the same time due to the fact that a minute atomic investigation of an organism must necessarily kill it. By regarding the wave function as universal, and regarding these aspects as complementary features of such a function, Bohr was able to propose that life is a purely physical phenomenon while at the same time denying that it could be explained in molecular terms. In this way he attempted to reconcile the intuition that there is more to life than atoms with a strictly materialistic point of view.

Of course, speculations of this kind have no basis in either sound reasoning or scientific evidence, but they do tend to lend an air of plausability to the assertion that the quantum theory, with all of its deficiences and bizarre features, must be accepted as final and absolute. In the hands of Bohr and his followers, the principle of complementarity became a tool for systematic question begging designed to protect the position of the quantum theory as the last word in fundamental knowledge. Any question tending to raise doubts about that theory was to be "renounced" as scientifically meaningless, and any apparant contradiction was to be explained as an example of complementarity. However, the quantum theory

53

itself was by no means to be renounced, even though it was new and only tested in limited circumstances (as has remained the case to this day.)

Splitting Universes

We briefly mention this theory only to indicate the irrational nature of the present scientific understanding of the world. According to this theory, the reduction of the wave packet corresponds to the splitting of the entire universe into multiple copies corresponding to the different alternative wave functions resulting from the reduction. Simultaneously, each person is supposed to split into corresponding multiple copies of himself which undergo different separate experiences in the different universes. These different copies never meet since the different universes are supposed to be totally isolated from one another. The splitting occurs continuously.[51]

Possibly this theory is not meant to be taken seriously, even though it has apparently been worked out in some detail. In any event, the very fact that such a theory has been officially proposed may be taken as symptomatic of the kind of knowledge which modern scientific theorization has been able to provide.

Quantum Logic

This is another category of proposals which we will only briefly consider. The basic idea is that logic is empirical, and that our traditional rules of logic are wrong.[52] It is argued that quantum mechanics demonstrates the need for a new system of logic, which is to be obtained by throwing out certain rules of logical deduction we now use. Once this is done, the dilemmas of quantum mechanics are supposed to disappear because there are no longer any lines of logical reasoning leading to them.

This theory has also been the subject of many articles and symposiums. It certainly seems to be a dubious method of resolving scientific problems—a method analogous to such things as the cure of insanity by prefrontal lobotomy. Apart from this, the proposal does not even address the fundamental problems involving the large scale ambiguity of the wave function which were discussed in the previous section.

This approach is significant in that it epitomizes a certain trend which has become very prominent in modern science. This is the tendency to confront basic questions by using one stratagem or another to declare them scientifically meaningless. We have already encountered this in our discussion of the Copenhagen interpretation, and it is also very prominent in modern biology, where it is now declared to be meaningless to ask what is the distinction between life and non-living matter. This tendency appears to be another manifestation of the basic desire, cherished by many scientists, that current scientific knowledge should be considered final and complete.

The physicist P. Dirac has expressed this basic tendency with the following words: "This, I think, is the kind of way in which we should try to develop our physical picture—to bring in ideas that make inconceivable the things we do not want to have."[53] The problem with this is: what if our ideas already render inconceivable things which are actually there? In that case, our desire to propound a closed system of thought will have led us down a blind alley which we can only escape by backing up and starting over again.

The Introduction of Consciousness into the Physical Picture

The view to be considered here is more directly relevant to our basic theme. It is based on the idea that the reduction of the wave function must be due to some definite cause which is not a product of the force laws summed up in the Schrödinger equation. The separate cause is considered to be consciousness, which must therefore be regarded as possessing an existence independent of those laws.

This view was first formulated by John von Neumann in his theory of the measurement process.[54] He considered the basic question of when and why process 1 must be invoked in the quantum mechanical analysis of an experimental measurement. He reasoned that the quantum theory describes an observer making a certain kind of measurement on an observed system. When the measurement is complete, the observer knows that the system has yielded a certain value for the measured parameter, and therefore process 1 must be applied to the wave function of the system in order to remove its ambiguity with respect to that parameter. In this situation one might suppose that the observer *caused* the effects of process 1 by his interaction with the observed system.

However, the postulate that quantum mechanics is complete implies that the observer can be included in the quantum mechani-

55

cal description. Von Neumann considered that the process of observation involves the transfer of information along a path from the observed object to the apparatus, from the apparatus to the observer's eye, from there to his optic nerve, and so on. He argued that one could draw a boundary line at any position along this path, and regard everything on one side of the line as the observed system, and everything on the other side as the observer. Thus, for example, one could say that the man was the observer and the experimental apparatus was the observed system. Or one could say that the apparatus plus the man's eyes were the observed system, and the man's brain plus optic nerve constituted the "observer."

No matter how the system is divided, process 1 must be invoked in order to assure that the observed system is unambiguous with respect to the features perceived by the observer. One can continue to suppose that this process occurs due to the influence of the observer, who always remains external to the observed system. However, if the boundary between observer and observed is move further and further into the brain, then one will eventually run out of observer.

Von Neumann argued that making the last step, in which the observer completely disappears, should not actually produce a fundamental change in the situation. He therefore proposed that when the entire physical system has been included on the observed side of the boundary, there still remains the "abstract ego" of the observer, which causes the effects of process 1 through its (or his) act of perception. This "abstract ego" is the non-physical consciousness of the observer.

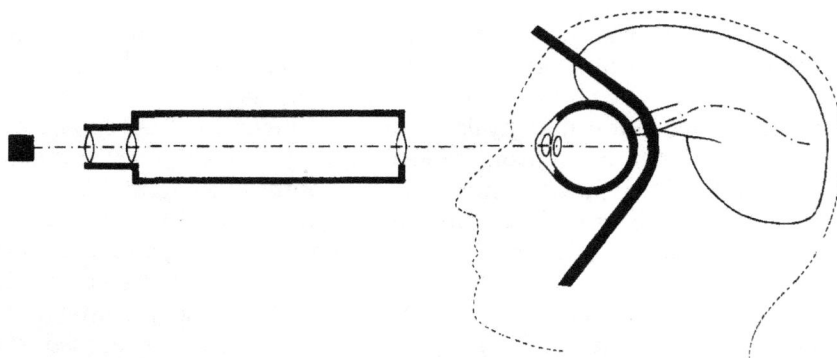

Figure 24. The division between the observer and the observed. The observer is to the right of the boundary line, and the observed system is to the left. The boundary line (or surface) can be shifted arbitrarily.

56

Von Neumann expressed his conclusion as follows: "We must always divide the world into two parts, the one being the observed system, the other the observer.... In particular we saw ... that the observer in this sense needs not to become identified with the body of the actual observer."[55]

This proposal answers two questions which had been evaded (or "renounced") in the Copenhagen interpretation. First, it provides an explanation of why the reduction of the wave function occurs. Second, it retains the observer as an external witness, while at the same time allowing for quantum mechanics to be universally applicable—at least to matter. It thus avoids the dilemma presented by a total system wave function which is assumed to encompass everything. If such a function represents the maximal possible knowledge an observer can have about a system, then where is that observer? Also, what relation does this function have with the consciousness of the persons whose bodies form part of the system?

However, von Neumann's proposal also suffers from the following basic drawbacks:

(1) It requires that changes in wave functions extending over arbitrarily large distances can be caused by the consciousness of one localized observer.

(2) The theory only deals with one observer. It does not indicate how many observers are to be introduced into the picture. (For this reason, von Neumann has sometimes been accused of solipsism.)

(3) It does not explain what is to be done about the need for continuous applications of process 1 which exists in some cases. (This was described in connection with Figure 21 in the last section.)

(4) It proposes that the effect of the observer's consciousness is *random*. This randomness is still taken to be causeless and inexplicable.

(5) No person has any conscious awareness of participating in or causing such random effects.

(6) This theory leaves unexplained the relation between the observer and the body of the observer.

In the section describing the Einstein, Podolsky, Rosen paradox, we saw that process 1 may involve sudden changes in a wave function over great distances. If this process is attributed to the consciousness of the observer, then it is hard to see how such a localized cause can produce such long range effects. This is especially significant if we consider that each individual person is normally aware only of the information provided by his immediate bodily senses, and has no knowledge of remote events.

Point number (6) basically summarizes the problems mentioned in this list. Under this heading we might also add the question of why certain material arrangements should be associated with observers, and not others. Human beings are certainly regarded as observers, but what about animals, plants, or stones? Where does the observer come from if he is more than just an arrangement of matter? How is the origin of observers to be understood in relation to the theory of evolution, which is solely concerned with matter acting under physical laws?

It is not surprising that no answers to these questions are to be found in von Neumann's theory. Some speculative suggestions as to how they might be answered have been made by E. Wigner, who is the principle advocate of von Neumann's point of view at the present time.[56] However, nothing very satisfactory can be said to resolve these difficulties—the quantum theory is simply not capable of dealing with such problems. Wigner himself admits that the present theory of quantum mechanics is not satisfactory and can be expected to be replaced by other theories in the future. He in fact proposes that "the present laws of physics are at least incomplete without a translation into terms of mental phenomena. More likely they are inaccurate, the inaccuracy increasing with the increase of the role which life plays in the phenomena considered."[57]

The basic problem of understanding the relation between matter and consciousness has been known in Western philosophy as the mind-body problem. Traditionally, philosophers have had great difficulty understanding how consciousness and matter interact. It is clear that the source or cause of consciousness has been left out of our present scientific theories. Yet, if we propose to assume the existence of non-physical conscious entities, then we are faced with the problem of explaining how they interact with matter which is obeying certain simple physical laws. Wigner has suggested that the resolution of this problem may involve the understanding that there are hitherto unknown natural laws associated specifically with life and consciousness.

V.

The Fallacy of Chance

Probability is the most important concept in modern science, especially as nobody has the slightest idea what it means.
—*Bertrand Russell*

In this section we shall briefly consider the significance of the concept of chance in our understanding of nature. In previous physical theories, chance was regarded as a measure of the ignorance of an observer about the underlying causes of observed events. However, everything that happened in nature was regarded as having a definite cause. Perhaps the most striking example of this is the theory of classical statistical mechanics, in which probabilities were used to describe the physical state of a gas or liquid containing an enormous number of atoms. It was necessary to introduce such probabilities because of the practical impossibility of actually acquiring knowledge about the individual motion of each atom. It was nonetheless taken for granted that each atom did have a specific, well defined position and velocity, and that the positions and velocities of all the atoms varied in accordance with the natural laws. The only absolute, or causeless features of the 19th century mechanistic world view were the initial conditions and the laws themselves.

In quantum mechanics, however, chance was taken to be causeless, or absolute. As we have seen, chance entered the theory through the rather arbitrary and awkward device known as the reduction of the wave packet.[58] As such, it plays the role of additional initial conditions, which are introduced repeatedly over the course of time to adjust the development of the wave function under the laws of motion. These adjustments had to be introduced in order to keep the physical picture consistent with the basic requirements of conscious awareness, but apart from this their role is not at all understood. We might also refer to these reductions as interim conditions.

The concept of chance in quantum mechanics is therefore quite different from the classical interpretation of chance as a measure of an observer's incomplete knowledge of underlying causes. Each individual random adjustment is regarded as an independent entity having no connection with anything else—it is

59

elemental and causeless. However, the concept of chance possesses an additional feature which conflicts with this absolute interpretation. This is the idea that "chance" refers to a spontaneous process which occurs in nature and obeys certain statistical laws. These laws specify a certain regularity in the overall, or average, behavior of random events. Apart from these regularities, however, the events are regarded as occurring freely in an uncontrolled and disorderly fashion.

This idea of a spontaneous process is actually essential to the theory of quantum mechanics. The average behavior of random events is described by the statistical laws in terms of certain numbers called probabilities. These probabilities are generated by the quantum theory. (Specifically, they are the probabilities mentioned in Figure 14 in the section on quantum mechanics.) The theory is linked with reality by the requirement that events in nature should satisfy the statistical laws, as expressed in terms of these probabilities. In fact, the theory is often confirmed by comparing its calculated probabilities with the observed average behavior of events. One of the main contributions of quantum mechanics to modern society has been the idea that the world is founded on the inherent lawlessness and disorder of a spontaneous random process.

Now, this poses a dilemma, for it implies that the overall arrangement of the reductions, or interim conditions, must be determined by the previous development of the wave function. How is this to be reconciled with the idea that each individual reduction is causeless?

In order to clarify this, let us consider the example of tossing a coin. Suppose that a coin is tossed repeatedly, and that after each toss a "1" or a "0" is written down to record whether the coin came up as heads or tails. After 100 tosses one might obtain the following record.

$$10110011010010110101010110000110101010010110101011 \atop 01100101001101010110101000001101101011110110101001 \qquad (18)$$

One would normally expect to find roughly half of these digits to be ones and half to be zeros, but one would expect to find no further recognizable pattern in the sequence. This situation is commonly described by saying that the outcome of each toss is due to chance, and that the probability of getting heads on any particular toss is ½.

It is expected that as the number of tosses increases we will find that

$$\frac{\text{The number of heads}}{\text{The number of tosses}} \rightarrow \text{½} = \text{The probability of heads} \qquad (19)$$

Here the arrow means that the quantity on the left approaches closer and closer to ½ as the number of tosses is increased. This is an example of a statistical law, known as the law of large numbers. The idea that the sequence (18) is disorderly can be expressed by means of similar statements. For example, suppose that X represents a particular pattern of N heads and tails, such as X = 1011001 for N = 7. Then we would expect to find that

$$\frac{\text{Times that N tosses in a row yield X}}{\text{The number of tosses}} \rightarrow \frac{1}{2^N}, \qquad (20)$$

as the number of tosses increases. This indicates that no one pattern is more likely to be found in the sequence of heads and tails than any other.

In quantum mechanics we might expect to find a sequence such as (18) occurring by "absolute chance." This would be the case, for example, if the sequence were determined in some appropriate way by the decay of radioactive atoms. The probabilities entering into (19) and (20) would then be determined by the quantum mechanical wave function. Although a large class of sequences will always exist which satisfies statements such as (19) and (20) within some degree of approximation, there will also be a large class which fails to satisfy them. To require a *causeless* sequence to avoid this class is contradictory, for it implies the existence of some active process by which the sequence is selected. This process would have to be added as an extra element to the structure of quantum mechanics as we have described it thus far.

The common notion holds that there is indeed such a process, and the existence of natural sequences which adhere to statistical laws is taken as proof of this. This is expressed in the following words in a standard textbook on probability theory: "The fact that in a number of instances the relative frequency of random events in a large number of trials is almost constant compels us to presume the existence of certain laws, independent of the experimenter, that govern the course of these phenomena and that

61

manifest themselves in the near constancy of the relative frequen-cy."[59] The words, "independent of the experimenter," indicate that these are natural laws. Let us consider what can be said about these laws. Can they be clearly defined in mathematical terms describing their functioning and their influence in nature, or must they remain a matter of vague intuition? Clearly the former must be the case if we are to claim that we understand matter and its laws of transformation.

Many attempts have been made to formulate a theory of probability in which chance is viewed as an acausal natural process governed by statistical regularities. Yet, even though an extensive literature has been devoted to this problem, no theory of this type has yet been devised which has been either practically applicable or free of serious inconsistencies. In particular, it has never been possible to assign a clear meaning to statements such as (19) and (20). A comprehensive evaluation of these theories has been given by the mathematician T.L. Fine.[60] We will not try to give a de-tailed account here of the many intricate arguments and counter-arguments which are involved in this subject matter. We will simply note the author's conclusion that

> The many difficulties encountered in attempts to understand and apply present day theories of probability suggest the need of a new perspective. *Conceivably, probability is not possible.* A careful sift-ing of our intuitive expectations and requirements for a theory of probability might reveal that they are unfulfillable or even logically inconsistent. Perhaps the Gordian knot, whose strands we have been examining, is best cut.[61] [Author's italics.]

Thus, the dilemma posed by attributing causeless chance to nature has not proven to be amenable to solution; the laws of chance needed to supplement the quantum mechanical laws are not available. However, the dilemma can be avoided if we give up the idea of absolute chance and return to the basic viewpoint that all material phenomena have definite causes.

It turns out that the "laws" described in (19) and (20) can be seen to follow as direct consequences of the assumption that the causes underlying the sequence of heads and tails are very *com-plex*. (This is, in fact, the common explanation of the unpredict-ability of the results of tossing a coin.) This is argued in formal mathematical terms in Fine's book.[62] It follows that we are not actually justified in supposing that a sequence is "lawless" or causeless simply because it obeys statistical rules such as (19) and

(20). On the contrary, this may be taken as an indication that the basic causes lying behind the sequence are inherently complicated and cannot be reduced to a simple scheme. The "laws" mentioned by Gnedenko can then be understood simply as corollaries of the causal laws of nature, rather than as the regulating principles of some mysterious "random process."

This is a very important point if we consider that the practical empirical basis for the assertion that the laws of nature are statistical has been the existence of observable sequences, such as (18), which satisfy statistical laws. Let us examine a number of different sequences which can be observed in nature:

(1) The sequence of clicks produced by a Geiger counter.
(2) The sequence of tosses of a coin.
(3) The works of Shakespeare, regarded as a sequence of letters.
(4) The sequence of sounds produced by several phone conversations which are garbled together so as to be heard simultaneously.
(5) The DNA base sequence for the chromosomes of a human being.

All of these sequences may be expected to satisfy statistical laws. In case (3), for example, one could calculate probabilities for the appearance of the various letters, and would find that "e" has the greatest probability. However, this does not mean that these sequences have come about by "chance." Certain of these sequences, such as (3), are filled with subtle meaning, and others, such as (4), are derived from meaningful sequences even though they are not meaningful themselves. It is thus reasonable to suppose that their adherence to certain statistical laws is due to the great complexity of causes lying behind them.

Case (5) is of particular interest. The biochemist Jacques Monod has reported that the amino acid sequences of large organic protein molecules satisfy the various statistical laws, as determined by extensive computer studies. He has concluded from this that these sequences are *random*. In his words, they represent "randomness caught on the wing, preserved, reproduced by the machinery of invariance and thus converted into order, rule, necessity. A *totally* blind process can by definition lead to anything; it can even lead to vision itself."[63]

Here we have an extreme example of the misunderstanding which we have been discussing. According to present biochemical understanding, a large protein molecule in a living cell performs very precise chemical operations involving coordinated interaction with many other molecular components of the cell. It stands to reason then that its structure should be complex but by no means chaotic or disorderly. This would account for its adherence to statistical laws, and we would not be forced to accept the rather contradictory view that this sophisticated structure was a product of "blind chance." Indeed, this is an instance where the marvelous precision and order of nature suggests the presence of natural causal principles extending far beyond the understanding of present day science.

One way of looking at the problems of quantum mechanics is to suppose that the theory is "under-determined," that is, lacking in instructions or laws which will determine the course of events in nature. The ad hoc introduction of "causeless chance" may be seen as an attempt to remedy this. However, the phenomena of chance can be better understood as the byproducts of complex causes than as the action of some mysterious "uncaused" sequences of events. This lends support to the idea that there exist many causal agencies or laws in nature unknown to modern science.

VI.

Quantum Laws versus Higher Order Laws

The possibility is always open that there may exist an unlimited variety of additional properties, qualities, entities, systems, levels, etc., to which apply correspondingly new kinds of laws of nature.

—D. Bohm

In Figure 25 we have indicated the complete Schrödinger equation for a system of electrons and atomic nuclei interacting according to the laws of electromagnetism. This may be regarded as an approximate statement of the ultimate causative principles underlying all the phenomena of chemistry and life, according to the current scientific view. Since this is the non-relativistic equation it is not perfectly in accord with the theoretical requirements of modern physics. It could be amended by the addition of various approximate relativistic correction terms (such as the Thomas precession, $(e/2Mc)\ \bar{\sigma} \cdot \bar{P} \times \bar{E}(\bar{Q})$.) The ultimate statement of final causes according to the physicists, however, would have to be the relativistic equation, as given by the theory of P. Dirac.[64]

We can thus see that it is difficult to write down a completely final version of the laws of nature as they are understood by modern science. The relativistic theory itself is plagued by conceptual and mathematical difficulties and cannot be regarded as a complete and consistent theory at the present time.[65] In addition, the equation presented here is based on the assumption that the atomic nuclei can be regarded as indivisible units. If this assumption were invalid, then we would have to take into account the vast array of mysterious and unexplained phenomena which have been discovered in the study of nuclear reactions at high energies. In that case we would not be able to give a definite mathematical description of the basic laws at all.

In view of this, it is strange that it is so commonly asserted that life can be understood completely in terms of the known laws of nature. The laws themselves are not actually known. What this may be understood to mean, however, is that life involves only chemical phenomena which occur within a certain range of energies. In this range the equation of Figure 25 is regarded as being a close enough approximation to the actual, though unknown, laws

65

$$\text{(a)} \quad H\Psi = i\hbar \frac{\partial}{\partial t} \Psi$$

$$\text{(b)} \quad H =$$

$$\sum_n \frac{-\hbar^2 c^2 \frac{\partial^2}{\partial q_n^2} + \eta_n^2 q_n^2}{2} + \sum_k \frac{-\hbar^2 \nabla_k^2}{2m_k}$$

$$+ \sum_k \frac{i\hbar e_k}{m_k c} \overline{A}(\overline{Q}_k) \cdot \nabla_k + \sum_k \frac{e_k^2}{m_k c^2} |\overline{A}(\overline{Q}_k)|^2$$

$$- \sum_k \frac{e_k}{2m_k c} \overline{\sigma}_k \cdot \nabla_k \times \overline{A}(\overline{Q}_k) + \sum_{i>j} \frac{e_i e_j - G m_i m_j}{|\overline{Q}_i - \overline{Q}_j|}$$

$$\overline{A} = \sum q_n \overline{A}_n$$

Figure 25. Quantum mechanical laws underlying chemistry.[66] Is this all that you are?

of nature to give a complete account of all of the phenomena of life.

Let us therefore consider some of the features of this equation. It has two features which are particularly striking:

(1) It is composed of sums of terms, each of which has a very simple format.
(2) It has been exactly solved only for a system of two particles. For larger numbers, only broad statistical conclusions have been drawn about the solution to the equation.

The first of these features simply reflects the fact that this equation is based on a very small number of extremely simple principles. Owing to the limitations of human reasoning power, this kind of simple structure is bound to be a feature of any equation which one might hope to seriously consider as a description of natural law. If we examine the equations in Figures 3 and 7 and Figures 9 and 25, we can see that these principles have been very slowly added and modified as different scientific theories have been proposed and rejected. One might well wonder how long a period of slow and painful development and revision would be necessary in order for scientists to develop equations based, say, on 20, or 100, or 1000 basic mathematical relations. It certainly has not been proven that a finite mathematical expression is sufficient to describe the laws of nature. Indeed, it is quite conceivable in principle that these laws cannot be approached even by an unlimited amount of mathematical information.

Essentially, this equation, like its predecessor equations, bases everything on sums of simple pushes and pulls. How is it that all of the phenomena of life could be due to such simple causes? Even if we neglect the phenomenon of consciousness, how are we to understand that chaotically distributed atoms and molecules have gotten together to form a world of incredibly intricate automatons, complete with eyes, brains, and molecular information storage systems? Why wouldn't they just keep on pushing and pulling in chaotic disarray? It is not sufficient to invoke the magical phrase, "natural selection," here. What we would like to understand is why pushes and pulls should be expected to combine in such a way as to produce such selection.[67]

Before one can even consider these questions of origins, there is the more immediate question of whether or not the equation of Figure 25 will describe the functioning of a living organism for *any* length of time. If one could determine the initial Ψ for a human body, and then compute the development of Ψ with time, what would one find? For all that we know, the body might immediately disintegrate, or even explode! We certainly have no basis for supposing that it will compose symphonies or create quantum mechinics through the mutual oscillation of myriads of pushes and pulls.

The difficulty mentioned in point (2) imposes a practically impenetrable barrier in the path of all attempts to demonstrate that equations such as this determine the functioning of living organisms. Even for as simple a system as the diatomic hydrogen molecule, H_2, the Schrödinger equation can only be solved by means of exceedingly difficult methods of approximation. Indeed, the solution has never been completely carried out. If we reflect that the human brain alone is estimated to contain some 100 billion nerve cells, we can see that there is no scientific basis for the claim that life is simply a product of the known physical laws. This conviction is simply a matter of blind faith.

The theory of quantum mechanics has in fact been tested experimentally only in very restricted situations. The physicist E. Wigner has pointed this out as an explanation of why discussions of the fundamental problems of quantum mechanics have not had much impact on the practical experimental applications of the theory. According to him, the reason for this is that "we hardly ever use quantum mechanics in the fashion we use classical mechanics: to predict events. Rather, we use it, as a rule, either to determine material constants, or the possible value of essentially only one observable, the energy. . . . It is perhaps disappointing that so few other conclusions of the general microscopic theory are commonly put to experimental test."[68]

We might, therefore, compare the assumption that quantum theory is universal with the conclusions reached by the scientific ant depicted in Figure 26. This ant is situated on the ceiling of a room, and its senses provide it with information about its immediate vicinity. After much observation and deliberation, it has developed a very elegant and concise theory of reality which it finds to be in excellent agreement with experiment (Figure 27). The ant has concluded that this theory is universal and complete.

Wigner has also made an interesting observation as to why scientists have so consistently tended to claim that the theory of the moment is a complete description of final principles. "If one admitted anything like the statement that the laws we study in physics and chemistry are limiting laws, . . . , we could not devote ourselves to our study as wholeheartedly as we have to in order to recognize any new regularity in nature. The regularity which we are trying to track down must appear as the all important regularity if we are to pursue it with sufficient devotion to be successful."[69]

Direct Evidence that the Quantum Theory Does Not Account for the Phenomena of Life

Owing to this determination to regard themselves as all-knowing, scientists have been willing in the past to relinquish a theory only when overpowering evidence was presented which disproved it. We may anticipate that this will also prove to be the case with the present dominant theories. The difficulties described thus far in the theory of quantum mechanics certainly indicate

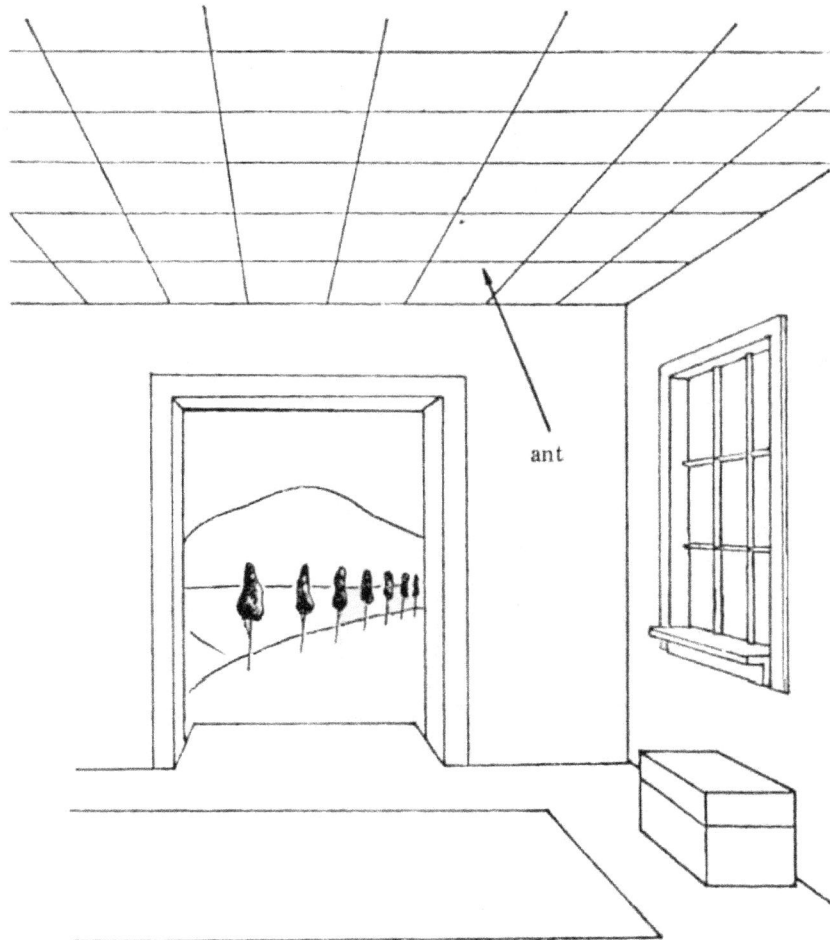

Figure 26. The scientific ant.

that the theory is incomplete in many respects. This would lead us to suspect that there should exist readily observable natural phenomena which directly violate the numerical predictions of the theory. In this subsection we will consider two lines of evidence of this type.

First, let us consider the phenomenon of biological transmutations. The French scientist C. L. Kervran has reported on extensive experiments showing that various chemical elements are created or destroyed within the bodies of living organisms.[70] He proposed that this could only be explained by

69

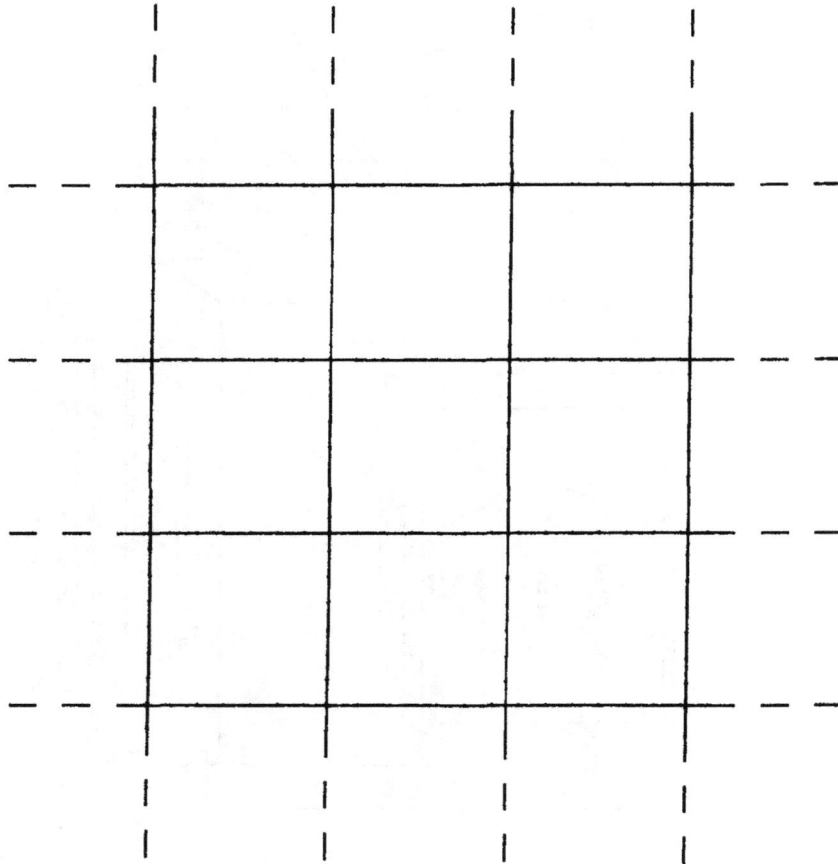

Figure 27. The ant's universal theory of reality.

supposing that nuclear reactions such as Na + O → K, Si + C → Ca, or Ca → Mg + O, could occur on a regular basis within the organisms as a feature of their normal metabolism. We mention this rather remarkable finding because it seems to have been based on careful experimental measurements and to have been corroborated by investigators other than Kervran himself.[71]

This result immediately implies that the phenomena of life are not simply limited to the realm of chemistry. In addition, it indicates the need for a revision in the basic laws of physics. Since atomic nuclei possess a positive electric charge, it follows from present theory that they strongly repel one another at close dis-

70

tances and cannot be brought together without the application of a considerable amount of energy. The problem of explaining how such nuclei could be combined or broken down in a systematic way within living cells should provide a considerable challenge to the theory of quantum mechanics.

The second example of contradictory evidence involves the phenomenon of self-reproduction. Wigner has presented an argument indicating that it is very doubtful that the equations of quantum mechanics can account for the capacity of living beings to reproduce themselves.[72] Here is a brief summary of his argument.

Consider a quantity of matter about the size of some living organism, such as an amoeba. If this matter is presumed to be within a certain range of energies, then it can be described by one of N different quantum mechanical states, or wave functions. This N will be an enormous number and will correspond to the total number of different configurations that the matter can assume within that range of energies.[73]

Let us suppose that out of these N states there are n "living" states, and that out of these n states there are m "viable" living states having the capacity to reproduce themselves in a favorable environment. Formally, we assume:

(a) For each viable living state there is at least one favorable environment. Taken together, this living state plus the environment will transform into a new state consisting of two living states plus the remainder of the environment. This transformation will take place in a definite time according to the Schrödinger equation for the system.[74]

Wigner showed that when this assumption is translated into quantum mechanical terms, it implies that a certain fixed set of linear equations must be satisfied. He observed that for these equations,

(b) $$\frac{\text{The number of unknowns}}{\text{The number of equations}} \approx \frac{n}{m}\left(\frac{n}{N}\right)^2.$$

These equations are determined mathematically by the quantum mechanical laws of the system and represent another way of expressing these laws.

It stands to reason that the number of viable living states should be nearly as great as the number of living states and that

71

the number of living states should be very small compared with the totality of states. Thus we should expect that

(c)
$$\frac{m}{n} \gg \frac{n}{N}$$

However, this implies that the number of linear equations which must be satisfied is vastly greater than the number of available unknowns. Wigner pointed out that this makes it very unlikely that the equations can be solved. Linear equations are generally expected to possess a solution only when there are at least as many unknowns as equations.

It therefore follows that the theory of quantum mechanics cannot be expected to account for the reproduction of living entities. The argument is not perfectly rigorous because the coefficients of the linear equations might just happen to line up in such a way that all of the equations could be simultaneously satisfied by the relatively small collection of unknowns. However, this would not be expected to occur unless there were some specific compelling reasons for it. It is certainly the responsibility of those who advocate the universality of present scientific knowledge to provide such reasons.

We should note in passing that the molecular model of cellular reproduction described by Crick and Watson has never been formulated exactly in terms of a definite system of natural laws. Indeed, many steps have been left unspecified in this model. It is only assumed without proof that the sequence of events entailed by this model correspond to the predictions of the "known laws of physics."

Higher Order Laws

The many limitations, contradictions, and bizarre features of quantum theory have led a number of scientists, such as Albert Einstein, to seek an entirely new theory.[75] Generally, these scientists have proposed that the quantum theory is incomplete, and that this incompleteness can be remedied by finding new variables which can be tied together by deterministic equations to form a theory of the 19th century type. Such proposals are generally referred to as hidden variable theories.

We shall not discuss these proposals in much detail since none of them have yet gained widespread acceptance. They have been the subject of much controversy, with some authors arguing that they are mathematically impossible in the context of modern experimental findings, and others in turn arguing that these authors are mistaken.[76]

Some of these scientists, however, have gone beyond the conception of developing a new mechanistic theory, and recognized that nature is not necessarily subject to description by a finite collection of mathematical laws. Thus, D. Bohm has stated that no matter how far our knowledge of nature progresses, we may always expect to find new laws, qualities, and properties which "lead to effects which are not small in relation to those following from previously known properties, qualities, and laws."[77] In other words, although we may express a limited and approximate understanding of nature by means of a finite system of laws, we may expect that any such system—no matter how elaborate—must fail to describe, or even touch upon, many significant natural phenomena. Such a view is quite plausible if we consider the incredible complexity and organization of living organisms and also the apparent incompatibility of consciousness with numerical description.

It is also supported by the many discoveries in pure mathematics of problems which can only be solved with the aid of unlimited systems of basic laws.[78] Perhaps the most famous example of this is Gödel's incompleteness theorem, which establishes that the theorems of arithmetic can only be established by an unlimited (non-recursive) axiom system. If this is true of mathematics, then it should not seem surprising that it might be true of nature also.

We would therefore like to introduce the concept of *higher order laws.* By the term "higher order" we shall refer to one of the following set of progressively stronger properties:

(1) The laws cannot be reduced to the known laws of physics.
(2) They can be expressed mathematically only by very elaborate expressions.
(3) They cannot be expressed mathematically at all, and relate to entities which are not amenable to numerical description.

From our investigation of quantum theory we can see that the presently known laws of physics do not give an adequate description of nature. At the very least, higher order laws of type (1) or type (2) must be considered. In the next section we shall describe how consciousness can most naturally be understood in terms of higher order laws of type (3).

VII.

The Laws of Consciousness and Matter

The absolute controller is acting within the heart, and is directing the wanderings of all living entities, who are seated as on a machine, made of the material energy.

—*Bhagavad-gītā*

In this review of various scientific theories we have been led to the conclusion that there are fundamental defects and limitations in man's present scientific understanding of matter and its laws. As we have seen, many theories have been proposed and rejected in the past, and this process still goes on. We have seen that the present dominant theory of quantum mechanics suffers from serious defects in its fundamental structure which can only be partially remedied by artificial, stop-gap measures. Its claim to universal applicability is unconfirmed, and there exists evidence which directly contradicts this claim. We can thus conclude that scientists do not know what matter is nor how it is acting.

The basic philosophical presupposition of modern science is that all the effects of nature are the consequences of a few simple laws capable of mathematical expression. Our thesis is that this presupposition has by no means been established. Indeed, as stated by the physicist D. Bohm in a discussion of this point, "the historical development of physics has not confirmed the basic assumptions of this philosophy, but rather has continually contradicted them."[79] In this section we would like to explore some of the possibilities that arise if we discard this philosophical assumption and suppose instead that there may be no limit to the variety of natural laws and entities. In particular, we would like to consider the role of consciousness as a phenomenon of nature.

The Irreducible Character of Consciousness

Throughout its struggles with the nature of matter, modern science has neglected consciousness almost completely, even though this phenomenon is the most primary feature of our existence as living beings. Indeed, the very existence of consciousness has proven to be a great embarrassment to the theoreticians of quantum mechanics. Some physicists, such as Niels Bohr, have been content to ignore, or "renounce," the very question of understanding consciousness. Others, such as von Neumann and Wigner, have recognized that consciousness lies outside the domain of their theories, but must be taken into account if a true understanding of nature is to be reached. However, they have not been able to introduce consciousness into their theoretical picture in a satisfactory way.

We would like to suggest that consciousness is a feature of reality which is incapable in principle of being adequately described in numerical terms. In our review of different mathematical descriptions of nature we have dealt with essentially two types of theories:

(1) In the first type there is a one to one correspondence between certain sets of numbers and the "fundamental elements" of reality. In such a theory every existing feature of nature can in principle be mathematically represented, for every feature must be a combination of fundamental elements. The classical nineteenth century theories were generally of this type.

(2) In the second type there is a correlation, which may be statistical, between certain numbers in the theory and the quantitative results of experimental measurements. It is not possible, however, to pin down the underlying reality which gives rise to these results. The calculations of the theory are simply supposed to predict the results to the greatest possible extent, and their structure cannot be thought of in terms of underlying entities or phenomena. The Ptolomaic astronomical system is an example of this kind of theory.

Generally, theories are interpreted as being of the first type. However, many people have felt impelled to regard quantum mechanics as a theory of the second category. (This is the option of admitting the wave function to be an incomplete description of nature, and simply accepting the whole theory as a set of approximate calculations.) Maxwell's electromagnetic theory can also be viewed in this way, as we pointed out in section II.

A theory of the second type may predict bodily movements or the electrical potentials of neurons in the brain very accurately (although we should stress that no such predictions have actually been made from any existing theory). However, it can give us no insight at all about how or why conscious awareness is associated

75

with such phenomena. From a calculated list of numbers corresponding to some physical behavior, what can we say about the awareness that may or may not have been associated with that behavior? It remains a complete mystery.

In a theory of the first type we are confronted with a picture of the world as a composite of many simple, elemental entities. As we have noted before, there is no reason to suppose that conscious awareness will exist just because many of these entities are juxtaposed in a certain pattern. Each simple, thoroughly insentient entity interacts with the others by some simple mechanical rule. At any one time each entity is changing (in position, orientation, amplitude, spin, or whatever) in a simple, thoroughly insentient way depending on this rule. No entity "knows" in any sense what the others are doing. How then can clear conscious awareness exist as a consequence of the presence of many such entities in some spatial arrangement?

The essential motivating idea for supposing that consciousness can be "explained" in this way is the notion that conscious awareness somehow corresponds with physical behavior. We have touched on this point before. However, to further dispel this misleading conception, let us consider a particular form in which it often appears. This is the idea that a computer can be conscious if it simulates by calculation the appropriate physical events occurring in a person's brain.

The British mathematician A. M. Turing has advanced the argument that all of a person's behavior can be duplicated by a suitably programmed computer.[80] Of course, this is far from being actually demonstrated. But, if we were confronted with such a computer, which could talk with us and exhibit all of the symptoms of human personality, then we might indeed be tempted to suppose that it possessed human consciousness. The question is: would this supposition be justified?

In order to answer this, consider what is going on within the computer during its calculations. Within the computer's "memory" unit there is stored a list of numbers encoding instructions for simple logical and arithmetical operations. Part of this list might look as follows:

code number   meaning

10 4787 0648   "Add the number at location 4787 to the number at 0648."

76

| 02 0648 1246 | "If the number at 0648 is positive go to the step stored at 1246; otherwise, go to the next step." |
| 03 0648 1267 | "If the number at 0648 is negative go to the step stored at 1267; . . ." |

. . . etc. . . .

All that the computer is doing at any one time is mechanically (or electrically) carrying out the instruction corresponding to one of these code numbers. The total behavior of the computer is simply the net result of the execution of many of these instructions, one after the other.

Since only a few simple electrical interactions are taking place at one time, it is hard to see how the computer could be conscious. If the computer were slowed down (as is possible) so that each simple step was stretched out over several seconds, the pattern and sequence of the steps would remain the same. Since the behavioral output of the computer would be slower but otherwise the same, does it follow that the conscious awareness of the computer would simply be stretched out in time? If not, we would have to explain why executing the instructions at one speed would generate conscious awareness of the thoughts being simulated, while at another speed there would be no consciousness of these thoughts.

Also, changing the construction of the computer should presumably not affect its consciousness as long as it is programmed to carry out the same steps, for this assures that its behavior will exhibit the same pattern. Figure 28 illustrates one form in which a computer can be constructed. Here the computer instructions are used to set up a gigantic "game" which could be played step by step by a child. As the child carries out these steps, will the same consciousness of the simulated thoughts be manifested there—stretched out, perhaps, over several years? This hardly seems plausible, but otherwise how are we to judge which of many computers with equivalent programs will be conscious and which ones will not?

We would like to suggest then that consciousness must be due to some existing entity in nature that cannot be numerically described. In the remainder of this paper we would therefore like to explore the implications of the following assumption: let us suppose that there are primary, irreducible entities that possess conscious awareness. These can be thought of as *quanta of con-*

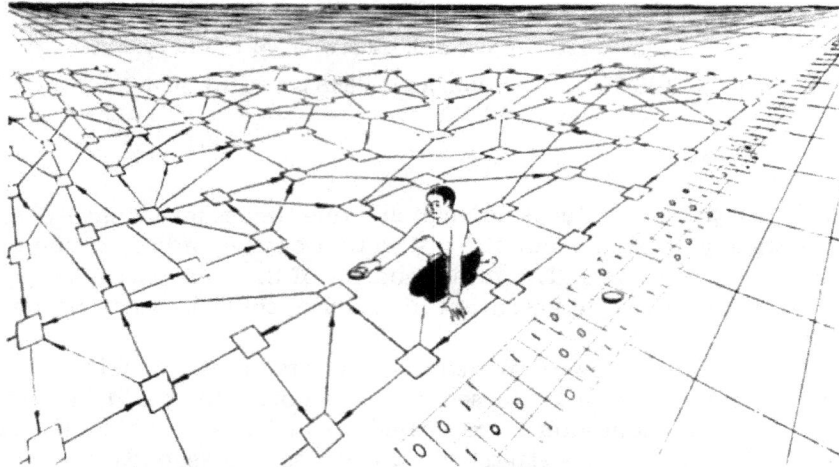

Figure 28. A computer in the form of board game. Suppose that the program, printed in steps on the squares of the board, is intended to recreate the consciousness of Joe Smith. As the game is played, will his consciousness be present there?

sciousness in analogy to electrons which, in standard physical theory, can be thought of as irreducible quanta of electricity. Each of these quanta of consciousness carries or possesses the individual awareness of a particular individual living being. We shall also use the term *atma* to designate one of these conscious entities.

Although we are assuming the conscious entity to be primary and irreducible, it is not simple like an electron. Whereas the electron is attributed simple properties, such as charge and spin, the quantum of consciousness must be capable of comprehending and appreciating very complex situations. For this reason we shall designate the quanta of consciousness collectively as the *superior energy*, in contrast with matter, which can be called the *inferior energy*.

Our proposal, then, is that each conscious living organism consists of a body composed of matter plus an *atma* which associates with this body and is conscious of the activities of the bodily senses. Many conscious entities must be invoked to account for the observed existence of many individual conscious beings. The content of the consciousness of a given *atma* interacting with matter will depend on the physical arrangement of that matter. For example, there will be consciousness of bodily sense perception if the senses are in working order, but there will not be such consciousness if they are not working. This state of affairs can be called *materially conditioned consciousness.*

By "matter" we mean the familiar object of study of physics and chemistry. Matter can be conceived of as an inherently insentient type of energy that can be transformed into many different temporary forms and configurations. The theory of quantum mechanics has left our understanding of matter rather "fuzzy" to say the least. It is capable to some extent of being described by various mathematical laws. However, it is evident that matter is still in many respects a mystery to modern science.

We should note that many of the features of our experience that are commonly called "mental" may be part of the materially determined content of consciousness, rather than part of the conscious entity, or *atma*, itself. This would include many different temporary features of our mental life such as the memory of words, and different skills and habits. However, the *atma* must be inherently capable of awareness and appreciation of these things, or they would simply go unnoticed.

Our basic proposition is that the quantum of consciousness cannot be described mathematically as a combination of insentient entities. The question naturally arises of whether or not the *atma* might be a combination of insentient, mathematically indescribable entities which fit together in a mathematically indescribable fashion. Actually, this is a rather cumbersome and intractable proposal. Since consciousness exists but cannot be numerically described, it follows that some kind of conscious entity must exist which cannot be understood in our familiar mathematical terms as a combination of simple, insentient entities. The simplest and most economical solution to this problem is to suppose that consciousness is an absolute feature of reality, and that the conscious entities are completely irreducible. This is much more in accord with Occam's razor, the principle of economy of thought, than is the introduction of many other completely mysterious irreducible entities.

Consider, for example, the hypothesis that consciousness is an "epiphenomenon" of brain activity.[81] This hypothesis is outlined in Figure 29. In this figure the laws governing matter are taken, for the sake of argument, to be of the classical 19th century type. The point of the epiphenomenon hypothesis is that certain material configurations, such as functioning human brains, "generate" consciousness, while others do not. The consciousness, on the other hand, is presumed to have no effect on matter (and there-

$$\text{Laws of matter:} \quad \frac{\partial p}{\partial t} j = F_1(p_1, \ldots, p_n; q_1, \ldots, q_n; j)$$

$$\frac{\partial q}{\partial t} j = F_2(p_1, \ldots, p_n; q_1, \ldots, q_n; j)$$

$$\text{Law generating consciousness:} \quad G(p_1, \ldots, p_n; q_1, \ldots, q_n) = \begin{cases} \text{``consciousness''} \\ \text{``unconsciousness''} \end{cases}$$

Figure 29. A model in which consciousness figures as an "epiphenomenon."

fore is not included among the arguments of the laws of matter, F_1 and F_2.)

In Figure 29, the law, G, which generates consciousness in this way is indicated. If G is a fundamental natural law then G must eternally exist in some sense. Yet G must also be mathematically indescribable since it generates consciousness, which is mathematically indescribable. G must somehow generate individualized, mathematically irreducible conscious awareness in association with each suitable brain. It is very hard to obtain a clear understanding of this mysterious G, or relate it, for example, to the theory that life has originated from matter by evolution—one might wonder what G was doing during all the time when no brains existed. The assumption of irreducible conscious entities certainly poses no more difficulties than the hypothesis of such a G.

However, this assumption has the interesting implication that the conscious self survives death. In fact, if we suppose that the *atma* tends to be associated with highly organized material bodies—and this certainly seems reasonable from an empirical standpoint—then the transmigration of the conscious self through a succession of bodies is implied. Even though these are very striking implications, we propose to adopt the assumption in this paper that the quantum of consciousness is completely irreducible. As we shall see, this assumption dovetails with a very concise and elegant world picture that is full of significant consequences and interesting avenues of further study.

The properties of the *atma* can be summarized as follows:

(1) The elemental carrier of consciousness.
(2) Exists in unlimited numbers.
(3) Cannot be created or destroyed (conservation principle).
(4) Tends to be associated with very complex bodies composed of material elements.

80

Universal Consciousness

Let us consider what sort of laws of interaction are involved with the quanta of consciousness. We shall approach this from the point of view of the basic question, "What is the nature of the absolute truth, or the final cause of all causes?" Here are a number of features characteristic of the concept of the absolute truth as we have encountered it thus far in normal science:

(1) The absolute truth exists, but is inconceivable to the human mind.
(2) It is all-pervading in space.
(3) It is invariant in time.
(4) It is the source and controller of all manifestations.
(5) It possesses an inherent unity.

As we have pointed out in previous sections, the natural laws have played the role of the final, absolute cause in the various mathematical theories of physics. Certainly the laws of all basic physical theories have shared properties (1), (2), and (3). Since the natural laws have always had to be supplemented by initial conditions (or even "interim" conditions!) they have not satisfied condition (4), however. Nonetheless, scientists have generally tried to minimize the ultimate importance of the initial conditions as much as possible by means of the idea of evolution. In this way they have tried to attribute property (4) to their systems of laws to the greatest possible extent.

Likewise, the laws of physics, inasmuch as they consist of a series of apparently unrelated mathematical expressions, do not satisfy (5). (See Figure 25.) However, the creators of theories have customarily tried to formulate their laws in such a way that they possess as much unity as possible. For example, the Hamiltonian formulation of Newton's laws in equations (1) and (2) of section II was considered to be a great accomplishment because of its unity and simplicity of form. Albert Einstein, in fact, devoted much of his efforts to the development of a "unified field theory" for physics because he felt that the ultimate cause underlying nature must be a harmonious unit rather than a disjointed mélange of unrelated things.

In this context, the simplest hypothesis that we could make about the absolute truth is the following: let us suppose that a "higher order law" governs the actions of both the inferior and superior energies, and that this higher order law satisfies conditions (1) through (5).

81

Since we are dealing with mathematically indescribable entities, this must be a higher order law of the third category mentioned in section IV. If the interaction of the *atma* with matter
proceeds according to such a law, then we might expect the behavior of matter to be more and more difficult to describe mathematically the more directly and intimately the matter is involved
with the activities of the *atma*. The more closely a mathematical
approximation predicts the measureable, external manifestations
of laws of this catagory, the more elaborate the approximation
must be. Therefore, while simple laws may be applicable to inanimate matter in standard laboratory situations, we may expect that
they will not suffice to describe the behavior of matter within the
bodies of living organisms. This corresponds to the view of the
physicist E. Wigner that "the present laws of physics are at least
incomplete without a translation into terms of mental phenomena.
More likely they are inaccurate, the inaccuracy increasing with the
increase in the role which life plays in the phenomena
considered."[82]

We have proposed that the "higher order law" we are considering should fully satisfy the criteria (1) through (5) for the absolute
truth. This means that this "law" must generate and control all
manifestations and also be highly unified. Certainly, unity in the
sense of simple equations is not possible here, for we must expect
even approximations to the material actions of this law to be
highly complex, the law itself being completely beyond mathematical description. It might seem very difficult to discuss or even
conceive of such an entity.

It turns out, however, that the simplest possible choice for our
"higher order law" satisfies points (1) through (5) perfectly. We
have already introduced consciousness as a primary, irreducible
feature of reality. The most economical hypothesis for the absolute causal principal is therefore the following: let us suppose that
the absolute truth is all-pervading, universal consciousness. We
shall designate this universal consciousness as *paramatma*.

Let us suppose that such things as desire, will, and purpose are
inherent aspects of the conscious entity, or *atma*. On this hypothesis, the interaction of the quanta of consciousness with matter
can be understood as follows: the *atma* situated within a particular
material body exhibits certain desires; these desires are perceived
by the *paramatma*, and in coordination with the desires of all
other living entities, the *paramatma* directs the material elements
of the body accordingly. This is possible since the *paramatma* is

the ultimate causative factor lying behind both the inferior and superior energies. Briefly, we are proposing that the interactive coupling between the *atma* and matter proceeds through consciousness.

In this picture, property (5) is satisfied since the *paramatma* is one conscious unit that perceives all things. We were already forced to posit an indivisible unit that can perceive a multiplicity of things when we introduced the *atma*, so this is not something qualitatively new. In this picture the individual *atmas* can be regarded as minute quantized parts of the *paramatma* which share its properties on a small scale.

We have also posited that the *paramatma* should satisfy property (4) of the absolute truth. By introducing laws of higher mathematical complexity—what to speak of laws transcending mathematical description—we have already been forced to depart from the spirit of evolutionary thinking. The idea of evolution is basically that complex form arises by simple processes from a situation where there was no such form. Here, however, we have already introduced absolute complexity. By proposing that the *paramatma*, or the ultimate causal agent, is the reservoir of all form and all processes, we are merely carrying this departure to its logical conclusion. This is also consistent with the requirement that each irreducible quantum of consciousness must possess the innate capacity to comprehend and appreciate complex material forms. It stands to reason that universal consciousness should have this same capacity on a universal scale. Thus, we propose that the *paramatma* possesses universal knowledge, and is thus able to manifest all phenomena without depending on chance or arbitrary initial conditions.

Consciousness and Knowledge

Thus far we have described the content of the consciousness of individual living entities as being limited to the perception of arrangements of matter. This content might be more or less elaborate depending on the condition of the body occupied by the conscious entity. However, it is natural to ask whether or not such a conscious entity can directly perceive either itself, other *atmas*, or the universal consciousness, *paramatma*.

This natural possibility opens up an entirely new avenue of practical investigation and is the most significant consequence of the view of reality which we have been outlining. Not only does it

83

open up a new line of inquiry into the nature of life and consciousness, but it also suggests a possible method of acquiring knowledge that is different in principle from the procedure of trial and guesswork employed in science: if the absolute cause underlying all phenomena is conscious, then one might hope to obtain knowledge from this source.

The existence of higher order laws and entities incapable of mathematical description makes it completely unrealistic to hope that comprehensive knowledge of life (or matter) can be attained solely by trial and error. This can be seen very easily if we reflect on the arduous struggles which led to the formulation of the physical laws illustrated in Figure 25. How long might it take to develop an equation involving 20 times as many terms and requiring as many revolutionary changes in fundamental conceptions of nature? Yet, these very considerations also point to an alternative source of knowledge.

Let us consider how the nature of the *atma* can be directly studied. An entity is normally studied in a scientific experiment by means of some procedure which isolates the entity in its pure and original form and eliminates extraneous influences. Such a procedure must take advantage of the particular distinguishing features of the entity and its laws of interaction.

For example, a beam of essentially pure "electricity" is generated by the apparatus known as a cathode ray tube, thus enabling many of the properties of electricity to be studied directly. This apparatus takes advantage of the basic property of electric charge and the laws of electrical interaction.

The study of the *atma* similarly requires some procedure for isolating it in its pure state. In our normal experience the *atma* is intimately bound up with matter by very powerful interactions, and therefore it is very difficult to discern its characteristic properties. In order to isolate the *atma* from the influences of material interaction it is necessary to take advantage of its basic distinguishing property—consciousness—and the agency—*paramatma*—governing its interactions. This requires the study of the relation between the individual conscious entity and the all-pervading absolute consciousness.

Thus, we cannot expect to study the *atma* using the familiar physical laws of interaction known to modern science. The *atma* cannot be expected to interact according to the laws of electromagnetism, and thus we could not expect to "see" one by means, say, of an electron microscope.

Rather, the isolation of the *atma* from the influences of matter requires procedures in which one's own self becomes the object of study and the primary features and characteristics of the self are invoked. Such systematic procedures have already been extensively studied, although they tend to be relatively unknown in the cultural tradition in which our modern scientific knowledge has been developed. Their basic principles are discussed very concisely, for example, in the ancient Sanskrit text known as the *Bhagavad-gītā*.[83]

This body of knowledge may be summarized as the science of self-realization, for it is logical, consistent, and systematically related with reproducible procedures of empirical observation. We have drawn upon a small part of this science in our presentation of the concepts of the *atma* and the absolute truth, or *paramatma*. We have tried to show the philosophical and scientific soundness of this system of ideas as a solution to the fundamental dilemmas faced by our present scientific picture of reality.

A detailed presentation of this science is beyond the scope of this paper. We would simply like to suggest that the approach to the study of consciousness which we have outlined here will lead to new insights in our understanding of both life and matter. On the other hand, we feel that the present scientific view of life as the electromagnetic interaction of certain molecules is quite wrong and will only lead to a frustrating dead end.

Appendix I

The Theory of Loinger, Daneri, and Prosperi

The goal of this theory is to solve the "problem of measurement" in quantum mechanics by eliminating the need to refer to feature (3)—the reduction of the wave packet. The basic idea is to show that only the laws of nature, as embodied in the Schrödinger equation, are needed in order to fully account for all phenomena. As such, this approach seeks to restore the integrity of physics by showing how everything can be explained in terms of the familiar concepts of physical forces without the introduction of arbitrary and mysterious adjustments, such as (3).[84]

It has been widely asserted that this theory has indeed accomplished its goal, and the authors themselves assert that "our theory constitutes an indispensable completion and a natural crowning of the basic structure of present day quantum mechanics."[85] However, the theory has also received serious criticism from many scientists.[86] We will not be able to discuss all of these criticisms here. This discussion will be devoted to one very basic point in which Daneri, Loinger, and Prosperi are in error.

In their theory it is proposed that owing to the essentially macroscopic (large scale) nature of the measuring apparatus, interactions occur during the process of measurement which, in effect, carry out the "reduction of the wave packet." These interactions are described by mathematical relationhips called ergodicity conditions.

The basic error in the theory derives from the misuse of the concept of a "mixture" or "density matrix." Let us briefly describe this concept. The result of the application of process 1 to a state, Ψ, can be summed up as,

(a) "The system is in the state, ψ_k, with probability $|c_k|^2$. (k = 1,2,3,...)"

Here $\psi_k = P_k\Psi/\|P_k\Psi\|$ and $c_k = \|P_k\Psi\|$. (See figure 14.) It is possible to correlate this statement in a one-to-one fashion with the mathematical expression,

(b) $$\sum_k |c_k|^2 \, P(\psi_k),$$

where $P(\psi_k)$ is called the projection operator associated with ψ_k.

The expression (b) is called a mixture or density matrix, and was first introduced by von Neumann.[87]

If the state of the system is Ψ, then it can also be referred to in terms of the mixture, $P(\Psi)$, since this corresponds to the statement,

(c) "The system is in the state, Ψ, with probability one (certainty)."

It is therefore possible to refer to both statements (a) and (c) by referring to the corresponding mixtures. In these terms, the reduction of the wave packet is formally equivalent to the transformation,

(d) $$P(\Psi) \quad \rightarrow \quad \sum_{k=1}^{\infty} \; |c_k|^2 \; P(\psi_k)$$

between mixtures.

What Daneri, Loinger, and Prosperi have argued is that, mathematically, the transformation of the wave function by Schrödinger's equation during a process of measurement is equivalent to the process (d), if one takes averages over a time interval. This is attributed to the very complex interactions occurring in the measuring apparatus as a consequence of the very large number of particles composing it, and is demonstrated by invoking the ergodicity conditions. They therefore conclude that Schrödinger's equation is sufficient to account for the reduction of the wave packet, and that this can be understood by properly taking into account the macroscopic size of the measuring apparatus.[88]

The error in their conclusion stems from the fact that after a "reduction of the wave packet" the system is not *in the mixture*,

$$\sum_{k} \; |c_k|^2 \; P(\psi_k).$$

Rather, it is in a particular state, ψ_k, which may be referred to by the mixture expression, $P(\psi_k)$, if we want to use this terminology. The "mixtures" appearing in (d) actually have two different interpretations. The system may be said to be *in the mixture* on the left, since this simply means that it is in the state, Ψ. However, the system cannot be said to be *in the mixture* on the right, since this expression refers to a statement of probabilities as to which state the system is actually in.

Factually, in the analysis presented by these authors the system is almost *always* in a state, $\Psi(t)$, which is ambiguous on a macroscopic scale. The mathematical approximation that

$$P(\Psi) \approx \sum_k |c_k|^2 P(\psi_k)$$

in an average sense, simply confirms this large scale ambiguity of Ψ. They have not shown how such ambiguities can be eliminated.

The statement that a system is *in* a mixture is quite common in the quantum mechanical literature and has apparently become a customary usage. On a macroscopic scale, it is clearly just as ambiguous to say that a system is *in* a mixture of two conflicting alternatives (such as day and night) as it is to say that it is in a state which is a superposition of states corresponding to those alternatives. Therefore the problem of ambiguity is not solved by saying that an ambiguous state, or wave function, has become equivalent to an equally ambiguous "mixture."

Another interpretation of mixtures is that they refer to ensembles of systems rather than to one particular system.[89] In this interpretation, statement (a) is taken to mean that N completely independent copies of the physical system are being considered, and that for each $k = 1,2,3,...$ we have $|c_k|^2 N$ of them in the state, ψ_k. However, we are interested in knowing what happens to one system. It clearly won't do at all to start out talking about one system, and then at some point change the subject to a discussion of a collection of systems. Indeed, this would require a repeated subdivision of the collection into larger and larger collections in those cases where the reduction of the wave packet must be repeatedly invoked. On the other hand, if we can only speak about the statistical behavior of large ensembles of systems without mentioning their constituents individually, then it follows that the quantum theory is drastically incomplete. Actually, these interpretations simply serve to obscure the fact that the problem of ambiguity in the quantum mechanical description of nature has not been solved.

This confusion over the meaning of mixtures also enters more subtly and indirectly into many discussions of quantum theory. This is illustrated, for example, in Rosenfeld's summary of the arguments of Daneri, Loinger, and Prosperi.[90] Without referring explicitly to mixtures, this author defines the reduction of the wave packet as the formal setting to zero of certain expressions, called correlation or interference terms. This, however, is simply the mathematical equivalent of the transformation in (d). The basic interpretation that process 1 means the replacement of a superposition of states by its corresponding mixture is still implicitly present. However, this simply begs the question of how and when an ambiguous state of affairs is to be converted into an unambiguous one in the quantum theory.

Appendix II

**Replacement of the Schrödinger Equation
by the Master Equation**

Here we shall briefly consider a type of equation, called the master equation, which can be used in place of the Schrödinger equation in some circumstances. This equation has the advantage of being conceptually simpler than the Schrödinger equation. It describes nature in terms of a Markov process which closely approximates the common, naive concept that molecular interactions proceed by actual random "quantum jumps" occurring one after another in individual molecules in ordinary time and space. It is also well suited to discussions of irreversible thermodynamic phenomena, such as the tendency of a large system of molecules to seek a state of equilibrium.

It is perhaps for these reasons that some biologists and biochemists have adopted this equation as an appropriate means of mathematically describing the metabolism and behavior of living organisms, and also their proposed origin from a "primordial soup" of simple chemical compounds.[91] Of course, this equation cannot be regarded as a complete description of nature; it belongs to that branch of physics, known as statistical mechanics, which entails the use of statistical assumptions to simplify the description of systems composed of very large numbers of interacting parts. It might be thought, however, that if living organisms are phenomena obeying the quantum laws, then this type of approximate equation might be useful for describing them.

We would simply like to point out here, however, that this equation cannot be expected to yield any information at all which specifically relates to living organisms. The reason for this is that the master equation describes nature in terms of an ensemble of wave functions with *random phases*.

This can be briefly explained as follows. In mathematics, classes of functions, ψ_n, are studied which have the property that any suitable function, Ψ, can be expressed as a linear superposition,

(a) $$\Psi = \sum_n a_n \psi_n,$$

of these functions. In particular, any quantum mechanical wave function for a physical system can be expressed as a superposition of energy eigenstates for that system. (The energy eigenstates are those wave functions for the system which have an unambiguous value for the energy.)

In this formula, the coefficients, a_n, are complex numbers which may be written as,

$$a_n = b_n \sqrt{P_n}, \quad |b_n| = 1.$$

(Each b_n is of the form $u + iv$, where $u^2 + v^2 = 1$.) The numbers, b_n, are called the *phases* of Ψ. The assumption of random phases means that we are not concerned with the values of the phases. We are only concerned with the entire class of wave functions obtained by varying the phases of Ψ over all possible values.

The physicist Van Hove describes the master equation as follows: "This equation, . . .

$$dP_n/dt = \sum_m (W_{nm}P_m - W_{mn}P_n),$$

involves the probability distribution P_n of the system over groups of eigenstates of the unperturbed hamiltonian H and expresses its irreversible time evolution under the action of the perturbation. It holds only to lowest order in the perturbation."[92] The equation describes how a *class* of wave functions with random phases changes with time, starting with the description of this class at some initial time.

Now, suppose that Ψ represents the body of some living organism. This means that $\Psi(\bar{q}_1, \ldots, \bar{q}_n)$ must be concentrated narrowly about the very specific configurations of atomic coordinates (the $\bar{q}_k$'s) corresponding to the highly intricate cellular structures making up the body of the organism. What happens, then, if we randomly change the phases of Ψ? The answer is that even small changes will completely wipe out any connection between Ψ and anything resembling a living organism. As a result, no *class* of Ψ's with random phases can tell us anything about living organisms.

In fact, the master equation was originally intended to describe such things as uniform gases, liquids, or solids in which no specific highly complex structures are considered. In Van Hove's derivation, the eigenstates, ψ_n, do not express any relation whatsoever between the different coordinates of the system.[93] Therefore, they do not individually describe molecular structures of even the simplest variety. It follows that Ψ, as a linear superposition of these states, can only single out such structures if its phases are very carefully adjusted.

91

Notes

1. Sherman, *Biology: A Human Approach*, p. 4.
2. Watson, *Molecular Biology of the Gene*, p. 67.
3. Watson, p. 105.
4. A.C. Bhaktivedanta Swami Prabhupāda, *Bhagavad-gītā As It Is*.
5. Heisenberg, *Physics and Beyond*, p. 112.
6. Motte, *Newton's Principia*, p. lxviii.
7. Einstein, Podolsky and Rosen, "Can Quantum Mechanical Description of Reality be Considered Complete?", p. 777.
8. Bell, *Men of Mathematics*, p. 172.
9. The term "natural law" can be used to refer to the abstract mental description of a regular pattern in the phenomena of nature, or to the reality underlying that pattern. (In the case of gravity this underlying reality might be described as "force" acting across empty space.) We will often use this term in the latter sense, especially when referring to laws which are posited as fundamental.
10. Koestler, *The Sleep Walkers*, p. 339.
11. A.C. Bhaktivedanta Swami Prabhupāda, *Krsna*, vol. III, p. 157.
12. Schrödinger, *What is Life? & Mind and Matter*, p. 117.
13. Wigner, "Two Kinds of Reality," p. 261.
14. Dyson, "Innovation in Physics," p. 76.
15. Planck, *Scientific Autobiography and Other Papers*, p. 33.
16. Messiah, *Quantum Mechanics*, vol. 2, pp. 1009–29. This reference illustrates some of the incredible mathematical maneuvering that these descriptions of "reality" involve.
17. In these equations, $\bar{A}_n$ (n = 1,2,3, . . .) is any complete orthonormal basis for the vector fields of divergence zero consisting of eigen-functions of the operator, ∇^2. (That is, $\nabla^2 \bar{A}_n = \eta_n^2 \bar{A}_n$ for each n.) There are many such sets of fields, and hence many equally valid models of this type. For consistency we have given the non-relativistic form of the theory.
18. The theory of relativity is another important theory which was devised early in this century. This theory is incorporated into quantum mechanics under the heading of "relativistic quantum mechanics."
19. Messiah, p. 876.
20. There may also be some additional variables, such as those describing the "spin" of the particles, which have no counterpart in the mechanical theory.
21. Technically, $|\psi(q_1,q_2)|^2$ is interpreted as a probability density function.
22. Nonetheless, wave functions are often depicted in this way in elementary texts. One secondary school presentation even featured fake "photographs" of what atomic wave functions would look like if one could see them. (See Cruetz, *The Science Teacher*, Sept. 1976, p. 30.)
23. Wigner, "Remarks on the Mind-Body Question," p. 286.

24. For simplicity we are neglecting the other variables which may appear in ψ.
25. Renniger, "Messungen ohne Störung des Meßobjekts," p. 420.
26. For general wave functions these probabilities are defined by integration of $|\psi|^2$ over volumes in the multi-dimensional domain of the function.
27. von Neumann, *Mathematical Foundations of Quantum Mechanics*, pp. 417–445.
28. There are actually many ways of defining process 1 if each a_k corresponds to many eigen-functions of A. (See von Neumann, p. 349.) The version chosen here seems to be in the best accord with the theory of measurement in quantum mechanics. This is another arbitrary feature of the theory.
29. This is treated in great detail in Fine, *Theories of Probability*. As an example: the standard Kolmogorof theory of probability does not actually define probability at all. It simply treats problems in the integration of exactly defined functions.
30. Wigner, "Two Kinds of Reality," p. 250.
31. Wilson, *Proc. Roy. Soc. London* (A) 87, 277 (1912).
32. This follows Mott, "The Wave Mechanics of α-Ray Tracks."
33. If ψ is the total wave function, then the alpha wave functions are F_{ijkl}, where

$$\psi = \underset{i\ j\ k\ l}{\Sigma\Sigma\Sigma\Sigma}\ F_{ijkl}\ \phi^1_i\ \phi^2_j\ \phi^3_k\ \phi^4_l,$$

and the ϕ's are the energy eigenstates of the target atoms.
34. If we add one gram of hydrogen gas to the system we will need some $2^{10^{28}}$ planes.
35. The planes which do not correspond to straight lines will not be very probable because, as shown in figure 19, the alpha waves tend to form narrow beams which travel in a straight line. This is because their wavelength is small compared with the effective size of the "target atoms." The wave length is inversely related with the energy of the "alpha particle" — $E = h/\lambda$. If the wave length of the alpha particle were longer, then the waves would tend to spread out, and planes representing many highly curved trails of exited atoms would be probable. This would correspond to the wave function of a lower energy particle.
36. Here is a more precise analysis of the wave function. Write this function as,

$$\psi = \underset{k}{\Sigma}\ F_k(\bar{R},t)\ X_k(\bar{Q},t),$$

where the X_k's represent histories of the behavior of the apparatus and observer. Take $\{X_k\}$ to be a complete orthonormal set and suppose, in accordance with our assumptions, that each X_k satisfies the Schrödinger equation for the system of observer plus apparatus (omitting the alpha particle.) Let $F_k(\bar{R},t)$ $(k = 1,2,3,...)$ be the alpha particle wave functions. Start at $t = 0$ with $\psi_0 = F_0 X_0$, representing un-ionized target atoms and no observation of a track. The Schrödinger equation becomes:

$$\left[i\hbar \frac{\partial}{\partial t} + \frac{\hbar^2}{2M} \nabla^2 \right] F_k(\bar{R},t) = \sum_n F_n(\bar{R},t) (V X_k, X_n),$$

where V is the potential of interaction between the alpha particle and the target atoms.

As time passes, $X_0(\bar{Q},t)$ should represent events in the absence of an "atomic decay." But, due to the term, $(V X_k, X_n)$, the alpha waves continually give weight to histories, $X_k(\bar{Q},t)$, differing from $X_0(\bar{Q},t)$ by the presence of ionized target atoms at time, t. These histories then develop differently from $X_0(\bar{Q},t)$ due to the chain of events leading from the formation of the ion trail, to the perception of the fog droplets, and so forth.

Thus, the wave function comes to simultaneously represent many different histories of events with nearly equal likelihood.

37. This kind of situation was discussed by Schrödinger, who considered a cat in a diabolical machine designed to kill it when triggered by the decay of a radioactive atom. The wave function becomes ambiguous as to whether the cat is dead or alive. This is widely known as the cat paradox. The version involving a human observer was discussed by E. Wigner in "Remarks on the Mind-Body Question."

38. Here we again encounter the problem that process 1 is not well defined for the macroscopic observation of variables such as position. By taking different definitions we can obtain different results. The basic conclusion, however, is that all macroscopic change in the system slows to a halt in the limiting case as the interval, Δt, between occurrences of process 1 goes to zero.

39. Even the well known "photograph" of atoms is simply a repeating pattern of phosphorescence on a screen which is produced by a certain apparatus. It is theoretically explained in terms of atoms.

40. See Einstein, Podolsky and Rosen, and also Thomsen, "Is Modern Physics for Real?"

41. For a discussion of this see Bohm, *Quantum Theory*, p. 614.

42. The electrons' wave function can be expressed as,

$$\psi = \sqrt{\tfrac{1}{2}} \left[(-+)_x - (+-)_x \right] = \sqrt{\tfrac{1}{2}} \left[(+-)_z - (-+)_z \right].$$

Here, $(-+)_x$ designates the wave function in which the first electron will definitely deflect in the minus x direction, and the second electron will deflect in the plus x direction. The other terms have an analogous meaning. If the observers measure deflection in the x direction, then ψ breaks down into $(-+)_x$ or $(+-)_x$. If he measures deflection in the z direction then ψ breaks down into $(-+)_z$ or $(+-)_z$. However, if ψ is first broken down into $(-+)_x$ or $(+-)_x$ and *then* the observers measure deflection in the z direction, the result turns out to be $(++)_z$, $(+-)_z$, $(--)_z$, or $(-+)_z$, which is not the same.

43. Dewitt, "Quantum Mechanics and Reality," p. 32.
44. See, for example, Bohm, "On Bohr's Views Concerning the Quantum Theory."
45. This is described in the discussion of figure 21 in the last section. It is also pointed out in Renniger, "Messungen ohne Störung des Meßobjekts."
46. Some scientists such as G. Ludwig have proposed that quantum mechanics is not valid for macroscopic systems. See Wigner, "Epistomological Perspective on Quantum Theory," p. 379.
47. Bohm, *Causality and Chance in Modern Physics*, pp. 91–94.
48. Heisenberg, "The Representation of Nature in Contemporary Physics," p. 107.
49. Bohr, *Atomic Physics and Human Knowledge*, p. 11.
50. Bohr, pp. 9–11.
51. Dewitt, "Quantum Physics and Reality."
52. Putnam, "Is Logic Empirical?"
53. Dirac, "The Evolution of the Physicist's Picture of Nature," p. 52.
54. von Neumann, *Mathematical Foundations of Quantum Mechanics*, chap. VI.
55. von Neumann, p. 420.
56. See Wigner, "Remarks on the Mind-Body Question."
57. Wigner, "Physics and the Explanation of Life," p. 44.
58. This also applies to the Born interpretation, in which $|\psi(\bar{Q})|^2 d\bar{Q}$ is taken as the probability of finding the particle coordinates in the arrangement, $\bar{Q}$. This probability arises in practice in a situation in which the reduction of the wave function singles out a particular coordinate arrangement—for example, in the observation of a scintillation screen.
59. Gnedenko, *Theory of Probability*, p. 55.
60. Fine, *Theories of Probability*.
61. Fine, p. 248.
62. Fine, p. 93.
63. Monod, *Chance and Necessity*, p. 98.
64. Messiah, *Quantum Mechanics*, vol. II, part 5.
65. Some of these difficulties are discussed in Wigner, "Epistomological Perspective on Quantum Theory," pp. 371–73.
66. This equation was basically obtained from the electromagnetic theory of the 19th century by the application of the correspondence rules. (See figure 8 of section II.) This would tend to attribute greater reality to the mechanical model of that theory (figure 7) than to the wave model (figure 4). Originally, the wave model was accepted as "real," whereas the radiation oscillators of the mechanical model appeared to be nothing more than rather arbitrary mathematical abstractions (Fourier coefficients, in fact.)

95

67. It is possible to argue, using information theory, that simple systems of laws in fact do not possess this kind of selective power. This is discussed in Thompson, *Demonstration by Information Theory that Life Cannot Arise from Matter.*

68. Wigner, "Epistomological Perspective on Quantum Theory," p. 374.

69. Wigner, "Remarks on the Mind-Body Question," p. 290.

70. Kervran, *Biological Transmutations.*

71. Baranger, "Do Plants Effect the Transmutation of Elements?"

72. Wigner, "The Probability of the Existence of a Self Reproducing Unit."

73. These different states are taken to be orthonormal.

74. Actually, Wigner allowed for the possibility that the living state plus environment would transform into an ambiguous superposition of states. He only required that each state in the superposition must contain at least two living states.

75. Schrödinger himself is said to have been dissatisfied with the way quantum mechanics developed. He once protested to Bohr that, "If we are going to stick to this damned quantum jumping, then I regret that I ever had anything to do with quantum theory." (Heisenberg, "The Development of the Interpretation of the Quantum Theory," p. 52.)

76. See, for example, Wigner, "On Hidden Variables and Quantum Mechanical Probabilities."

77. Bohm, *Causality and Chance in Modern Physics,* p. 133.

78. Jones, "Recursive Undecidability—An Exposition."

79. Bohm, p. 131.

80. Turing, "Computing Machinery and Intelligence."

81. See *The Encyclopedia of Philosophy,* vol. 5, p. 343.

82. Wigner, "Physics and the Explanation of Life," p. 44.

83. A. C. Bhaktivedanta Swami Prabhupada, *Bhagavad-gita As It Is.*

84. See Daneri, Loinger and Prosperi, "Further Remarks on the Relations Between Statistical Mechanics and Quantum Theory of Measurement," and "Quantum Theory of Measurement and Ergodicity Conditions." A summary of this theory is presented in Rosenfeld, "The Measuring Process in Quantum Mechanics."

85. Daneri, Loinger and Prosperi, "Further Remarks . . .", p. 127.

86. Some of these are:
Jauch, Wigner and Yanase, "Some Comments Concerning Measurement in Quantum Mechanics,"
Freundlich, "Mind, Matter, and Physicists,"
Zeh, "On the Interpretation of Measurement in Quantum Theory."

87. von Neumann, *Mathematical Foundations of Quantum Mechanics,* chap. IV.

88. They deal with an experiment involving a brief interaction between an atomic system and a macroscopic apparatus. After the interaction, the systems are uncoupled and in the state,

$$\Psi(t) = \sum_k c_k \; \phi_k(t) \; \Phi_k(t).$$

96

The mathematical essence of their argument is that,

$$P(\Psi(t)) \simeq \sum_{k} |c_k|^2 \; P(\phi_k(t)\Phi_k(t))$$

on the average over the time, t, and over the components of the "cells" into which they divide the macroscopic system.

89. This is discussed in Freundlich, "Mind, Matter, and Physicists."
90. Rosenfeld, "The Measuring Process in Quantum Mechanics."
91. See, for example, Fox, "A Non-Equilibrium Thermodynamical Analysis of the Origin of Life."
92. Van Hove, "The Approach to Equilibrium in Quantum Statistics," p. 441.
93. This is because ". . . H is assumed to give an essentially complete separation of variables." (Van Hove, p. 443.)

Bibliography

A. C. Bhaktivedanta Swami Prabhupāda. *Bhagavad-gītā As It Is.* New York: Collier Books, 1972.

A. C. Bhaktivedanta Swami Prabhupāda. *Kṛṣṇa,* Vol. III. Los Angeles: Bhaktivedanta Book Trust, 1970.

Baranger, P. "Do Plants Effect the Transmutation of Elements?" *Mother Earth, Journal of Soil Association,* April 1960, pp. 153-6.

Bohm, D. *Causality and Chance in Modern Physics.* van Nostrand, 1957.

Bohm, D. "On Bohr's Views Concerning the Quantum Theory." *Quantum Theory and Beyond,* ed. T. Bastin. Cambridge: Cambridge Univ. Press, 1971, pp. 33-40.

Bohm, D. *Quantum Theory.* New York: Prentice Hall, 1951.

Bohr, Niels. *Atomic Physics and Human Knowledge.* New York: John Wiley and Sons, 1958.

Bell, E.T. *Men of Mathematics.* New York: Simon and Schuster, 1965.

Daneri, A., Loinger, A., Prosperi, G.M. "Further Remarks on the Relations Between Statistical Mechanics and Quantum Theory of Measurement." *Il Nuovo Cimento,* Vol. 44 B, No. 1, 1966, pp. 119-28.

Daneri, A., Loinger, A., Prosperi, G.M. "Quantum Theory of Measurements and Ergodicity Conditions." *Nuclear Physics,* Vol. 33, 1962, pp. 297-319.

Dewitt, B.S. "Quantum Mechanics and Reality." *Physics Today,* Sept. 1970, pp. 30-35.

Dirac, P.A.M. "The Evolution of the Physicist's Picture of Nature." *Scientific American,* Vol. 208, No. 5, May 1963, pp. 45-53.

Dyson, F.J. "Innovation in Physics." *Scientific American,* Sept. 1958, pp. 75-82.

Edwards, P., ed. *The Encyclopedia of Philosophy,* Vol. 5. New York: Macmillan, 1967.

Einstein, A., Podolsky, B., Rosen N. "Can Quantum Mechanical Description of Reality be Considered Complete?" *Physical Review,* Vol. 47, May 1935, pp. 777-80.

Fine, T.L. *Theories of Probability.* New York and London: Academic Press, 1973.

Fox, R. "A Non-Equilibrium Thermodynamical Analysis of the Origin of Life." *Molecular Evolution,* ed. Rohlfing and Oparin. New York: Plenum Press, 1972.

Freundlich. "Mind, Matter, and Physicists." *Foundations of Physics,* Vol. 2, No. 2/3, 1972, pp. 129-47.

Gnedenko, B. *Theory of Probability,* 4th ed. New York: Chelsea, 1968.

Heisenberg, W. "The Development of the Interpretation of the Quantum Theory." *Niels Bohr and the Development of Physics,* ed. W. Pauli. New York: McGraw Hill, 1955.

Heisenberg, W. *Physics and Beyond.* New York: Harper and Row, 1971.

Heisenberg, W. "The Representation of Nature in Contemporary Physics." *Daedalus*, Vol. 87, No. 3, 1958, pp. 95-108.

Jauch, A.M., Wigner, E.P., Yanese, M.M. "Some Comments Concerning Measurement in Quantum Mechanics." *Il Nuovo Cimento*, Vol. 48, No. 1, Mar. 1967, pp. 144-51.

Jones, J.P. "Recursive Undecidability—An Exposition." *American Mathematical Monthly*, Sept. 1974, pp. 724-38.

Kervran, C.L. *Biological Transmutations*. Binghamton, New York: Swan House Publishing Co., 1972.

Koestler, A. *The Sleep Walkers*. New York: Macmillan, 1959.

Messiah, A. *Quantum Mechanics*, Vol. 2. Amsterdam: North Holland, 1961-62.

Monod, J. *Chance and Necessity*. New York: Alfred A. Knopf, 1971.

Mott, N.F. "The Wave Mechanics of α-Ray Tracks." *Proceedings of the Royal Society*, London, Vol. 126, 1929, pp. 79-84.

Newton's Principia, Motte, A., trans. New York: Daniel Adee, 1846.

Planck, M. *Scientific Autobiography and Other Papers*. Westport, Conn.: Greenwood Press, 1949.

Putnam, H. "Is Logic Empirical?" *Boston Studies in the Philosophy of Science*, Vol. V. Boston: Reidel Publishing Co., 1969.

Renninger, M. "Messungen ohne Störung des Meßobjekts." *Zeitschrift fur Physik*, Vol. 158, 1960, pp. 417-21.

Rosenfeld, L. "The Measuring Process in Quantum Mechanics." *Suppl. Progr. Theor. Phys.*, extra n., 1965, pp. 222-31.

Schrödinger, E. *What Is Life? & Mind and Matter*. Cambridge: Cambridge Univ. Press, 1967.

Sherman, I.W. & V.G. *Biology—A Human Approach*. New York: Oxford Univ. Press, 1975.

Thompson, R.L. *Demonstration By Information Theory that Life Cannot Arise from Matter*. Boston: Bhaktivedanta Institute, 1977.

Thomsen, D.E. "Is Modern Physics for Real?" *Science News*, Vol. 109, No. 21, May 22, 1976, pp. 332-3.

Turing, A.M. "Computing Machines and Intelligence." *Mind*, Vol. 59, 1950, pp. 433-60.

Van Hove, L. "The Approach to Equilibrium in Quantum Statistics." *Physica*, Vol. 23, 1957, pp. 441-80.

von Neumann, J. *Mathematical Foundations of Quantum Mechanics*. Princeton: Princeton Univ. Press, 1955.

Watson, J.D. *Molecular Biology of the Gene*, 2nd ed. Menlo Park, Calif.: W.A. Benjamin, Inc., 1970.

Wigner, E.P. "Epistomological Perspective on Quantum theory." *Contemporary Research in the Foundations and Philosophy of Quantum Theory*, ed. C.A. Hooker. Boston: Reidel Publishing Co., 1973.

Wigner, E.P. "On Hidden Variables and Quantum Mechanical Probabilities." *American Journal of Physics*, Vol. 38, No. 8, 1970, pp. 1005-9.

Wigner, E.P. "Physics and the Explanation of Life." *Foundations of Physics*, Vol. 1, No. 1 1970, pp. 35–45.

Wigner, E.P. "The Probability of the Existence of a Self Reproducing Unit." *The Logic of Personal Knowledge*. Routledge and Kegan Paul, 1961.

Wigner, E.P. "Remarks on the Mind-Body Question." *The Scientist Speculates*, ed. I.J. Good. New York: Basic Books Inc., 1963.

Wigner, E.P. "Two Kinds of Reality." *The Monist*, Vol. 48(2), 1964, pp. 248-63.

Wilson, C.T.R. *Proceedings of the Royal Society*, London(A), Vol. 87, 1912, p. 277.

Zeh, H.D. "On the Interpretation of Measurement in Quantum Theory." *Foundations of Physics*, Vol. 1, No. 1, 1970, pp. 69-76.

About the Author

Sadāputa dāsa (Richard Thompson) was born in Binghamton, New York, on February 4, 1947. In 1969 he earned his B.S. degree in mathematics from the State University of New York at Binghamton, and in 1970 he earned his M.A. in mathematics from Syracuse University. After receiving a National Science Fellowship in 1970, he completed his Ph.D. in mathematics at Cornell University in June of 1974, specializing in probability theory and statistical mechanics. His dissertation has been published as Memoir number 150 of the American Mathematical Society, "Equilibrium States on Thin Energy Shells."

Throughout his studies, the author was struck by the lack of any meaningful foundation to reality in modern scientific theories. His dissatisfaction with this culminated in 1970, when he studied the reduction of man to a Turing machine, a kind of abstract clockwork with a few moving parts. Surely, he felt, the truth must be something different from this. Consequently, he began to study many different philosophies, with a view to finding a practical route to higher knowledge.

In 1972 he discovered some of the books of His Divine Grace A.C. Bhaktivedanta Swami Prabhupāda in a book store in Ithaca, New York, and was struck by the beauty of their conceptions and the clarity of their presentation. Later he met the disciples of Śrīla Prabhupāda at the Rādhā-Krishna Temple in New York City. Here, he found, was a deeply meaningful philosophy capable of practical application in day-to-day life. He became formally initiated as Śrīla Prabhupāda's disciple in 1975 at the temple of Śrī Śrī Gaur-Nitai in Atlanta, Georgia. He is now a full-time member of Bhaktivedanta Institute for Higher Studies.

SECTION II

"LIFE COMES FROM LIFE" CONFERENCE DOCUMENTS

Section II includes the following items related to the conference:

(1) The title page from an early cooperative effort that would eventually become the first monograph.

(2) An announcement for the "Life Comes from Life" conference offering a brief outline of its objectives.

(3) A brochure titled, "Bhaktivedanta Institute Lecture Series," which presents a project mission statement along with an outline of programs, areas of research, and short biographies of the Bhaktivedanta Institute members speaking at the conference.

(4) Copies of a trifold invitation brochure originally printed on light yellow cardstock offering a description of conference objectives and a schedule of events.

(5) A list of speakers with a description of their professional qualifications at the time of the conference.

(6) A finalized schedule of events with the name and topic of each presenter.

(7) Text from an article reporting on the conference published in *The Statesman*, a widely distributed national newspaper headquartered in Calcutta. This account is taken from audio recordings of the article being read during a follow-up meeting to the conference.

(8) The final document was published in the December 1977 edition of *Back to Godhead*, "the magazine of the Hare Krishna Movement," reporting the events to ISKCON's congregation.

"WHAT IS MATTER AND WHAT IS LIFE?"

(first version)

by:

Svarupa Damodara Dasa (Dr. T.D. Singh)

Sadaputa Dasa (Dr. R. Thompson)

Madhava Dasa (Dr. M. Marchetti)

(Contribution from BHAKTIVEDANTA INSTITUTE)

9/16/76.

Copy to:
1) Sadaputa Prabhu
2) Madhava
3) Rupanuga
4) Ravindra Svarupa
5) Wolf

LIFE   COMES   FROM   LIFE

First International Scientific Conference

Sponsored by - BHAKTIVEDANTA   INSTITUTE, Vrindaban, Mathura

October 14-16, 1 9 7 7

A brief outline of the conference

The study of life and its origin forms the most fundamental
basis of all knowledge. The currently dominant scientific
theory claims that life is a coordinated chemical reaction
and originated from a "primordial chemical broth" under the
influence of chance and blind natural laws. Based on this
assumption, there have been many false reports of the chemical
origin of life from scientists around the world, for example ,
Fox in the U.S.A., Crick in the United Kingdom, Oparin in the
USSR and Krishna Bahadur in India.

The main emphasis of the conference will be focussed on the
fact that the above theory is scientifically unfounded and
philosophically unsound. A new scientific paradigm is
clearly needed to understand life, its value, meaning and
purpose.

The scientists of Bhaktivedanta Institute, the academic
division of the world-famous International Society for Krishna
Consciousness, propose that life is non-physical and non -
chemical. It possesses the quality of consciousness, and
obeys higher order, non-physical natural laws. It also
originates from a supremely conscious Being, Supreme Life or
God. Such an understanding is based on the description of
the atomic living entity, atma , and its relationship with
paramatma, Supreme Life, given in Bhagavad-gita.

The argument will be presented in the light of currently
available data of physical, chemical, and biological sciences.

Bhaktivedanta Institute Lecture Series

on

THE ORIGIN OF LIFE AND MATTER

BHAKTIVEDANTA INSTITUTE

Founder — Ācārya

His Divine Grace A.C. Bhaktivedanta Swami Prabhupāda

About Bhaktivedanta Institute

Bhaktivedanta Institute is a center for advanced study and research into the Vedic scientific knowledge concerning the nature of consciousness and the self. The Institute is the academic division of the International Society for Krishna Consciousness. It consists of a body of scientists and scholars who have recognized the unique value of the teachings of Krishna Consciousness brought to the West by His Divine Grace A. C. Bhaktivedanta Swami Prabhupāda. The main purpose of the Institute is to explore the implications of the Vedic knowledge as it bears on all features of human culture, and to present its findings in courses, lectures, monographs, books and a quarterly journal, *Sa-vijñānam*.

The Institute presents modern science and other fields of knowledge in the light of Vaiṣṇava philosophy and tradition, providing a new perspective on reality quite different from that of our modern educational systems. One reason for the increasing interest of modern intellectuals in Śrīla Prabhupāda's teachings is doubtlessly the growing awareness that in spite of great scientific and technological advancements, the real goal of human life has somehow been missed. The philosophy of Bhaktivedanta Institute provides a meaningful answer to this concern by proposing that life—not matter—is the basis of the world we perceive.

The central doctrine of modern science is that all phenomena, including those of life and consciousness, can be fully explained and understood by recourse to matter alone. The dictum that "life is a manifestation of matter" is, indeed, the ultimate rationale for the entire civilization of material aggrandizement. The Vedas, on the other hand, teach that conscious life is original, fundamental, and eternal. This is the essence of *Bhagavad-gītā*—*"ahaṁ sarvasya prabhavo mattaḥ sarvaṁ pravartate."* (10.8) On this fundamental and critical point, modern science and Vedic knowledge find themselves opposed.

Bhaktivedanta Institute is dedicated to disseminating this most fundamental knowledge throughout the world. The Institute is clearly demonstrating that the Vedic version is not a matter simply of "faith" or "belief", but is scientific in the strict sense of the term. Although many of its features may appear difficult to verify experimentally, others have direct implications concerning what we may expect to observe. Thus, this view should serve as a stimulating challenge to the truly scientific spirit to go beyond the very restrictive framework imposed on our scientific understanding of nature over the last two hundred years. Modern science began as an experiment to see how far nature could be explained without invoking God. But the purpose of Bhaktivedanta Institute is to introduce Vedic knowledge on a genuinely scientific basis for the first time in the history of this modern scientific age.

Program of Lectures on the Origin of Life and Matter

Today, science is dominated by the theory that "life" is a product of molecular interactions, and that all the different activities of life—for example, thinking, feeling and willing—are due to coordinated chemical reactions. Yet, this theory suffers both from serious internal inconsistencies and a fundamental inability to account for many of the most significant phenomena of life. A new scientific paradigm is clearly needed.

Bhaktivedanta Institute is sponsoring a lecture series on the nature and origin of life and matter. Institute members are available to present lectures on many different aspects of this fundamental subject, both for general audiences and for specialists in several fields of modern natural science. The basic theme for the

Structure of the protein, chymotrypsin. Can such precise and complex structures arise by chance and the action of molecular forces?

lectures is that life cannot be reduced to a combination of material elements. This is discussed in the light of modern scientific theories and the Vedic knowledge. The members of the Institute hope that these discussions will provide new and thought provoking insights into these important questions.

Lecture Topics

- The Fundamental Difference between Life and Matter

- Quantum Theory and Consciousness

- Higher Order Natural Laws—a New Scientific Paradigm

- Information Theory and the Origin of Complex Order

- Chemical Evolution: A Molecular Fairy Tale?

- Darwin's Theory in Retreat—New Trends in Geological Interpretation

- Genetic Engineering and Biomedical Ethics

- The Philosophy and Psychology of Science

- *Paramātmā* and the Process of Acquiring Knowledge

- A Scientific Basis for the Study of Life

Representation of a computer in the form of a game. Can a computer in this form generate a person's consciousness? If not, how is the role of consciousness in nature to be understood?

Some Highlights of the Lecture Series

- Molecular Biology
  The goal of scientific research is to find the absolute truth, or the cause of all phenomena, governing both life and matter. According to modern science, the ultimate cause is vaguely incorporated into certain physical laws—basically the laws of quantum mechanics. Conceptually, these laws involve only some pushes and pulls among particles. Although the theory of evolution asserts that these laws are sufficient in themselves to account for all the marvels of life, honest and intelligent scientists are beginning to realize that the theory is failing to explain many of life's subtle aspects—for instance, love, meaning and purpose. How can simple pushes and pulls be responsible for all the wonderful phenomena that we encounter in life?

  We would like to show that the recent announcement of Khorana's synthetic gene is not different from that of Wöhler's synthesis of urea in 1828, as far as our understanding of life is concerned. In other words, a complex molecule or a combination of such molecules will not account for the true nature of life.

  We propose that life is non-physical and non-chemical. It possesses consciousness and obeys higher order non-physical laws. This hypothesis can explain all the observed facts of life more scientifically than any of the previous theories.

- Information Theory
  A fundamental proposition of information theory states that the information content of a closed mathematical system cannot increase. In modern science, nature is described by means of mathematical models of low information content. The physical structures of living organisms, on the other hand, are of such complexity and diversity as to indicate a very high information content. From this it follows that these structures could not have arisen by means of the simple natural processes envisioned in the theory of evolution. An additional source of information is required. The implications of this analysis regarding the origin of life are discussed.

- Quantum theory
  Modern science has failed to account for consciousness, and in modern physics the existence of consciousness has given rise to serious paradoxes and contradictions. The quantum theory is discussed with special emphasis on the interpretations of von

Neumann and Daneri, Loinger, and Prosperi. It is concluded that a new theoretical understanding is required containing provisions for consciousness. Such an understanding may be based on the description of the atomic living entity, or *ātmā*, given in *Bhagavad-gītā*. This entails definite implications concerning the scientific study of life.

Literature

The members of the Institute are writing books and monographs, and publishing a journal, *Sa-vijñānam*, devoted to the study of scientific knowledge. Topics of particular interest are: the nature and origin of life, consciousness, the theory of evolution, the laws of nature, and the psychology and methodology of science. The following monographs are currently available:

1. *What is Matter and What is Life?*
       — by Svarūpa Dāmodara Dāsa

2. *Demonstration by Information Theory that Life Cannot Arise from Matter*
       — by Sadāputa Dāsa

3. *Consciousness and the Laws of Nature*
       — by Sadāputa Dāsa

Speakers

Dr. Thoudam Damodar Singh (Svarūpa Dāmodara Dāsa), GBC and Director of Bhaktivedanta Institute

Dr. Singh was born in Manipur, India, in 1941. He received his B.Sc. with First Class Chemistry Honors from Gauhati University, and his Master of Technology degree with First Class Honors from Calcutta University. He received his M.S. in Chemistry from Canisius College of Buffalo, New York, and in 1974 completed his Ph.D. in Physical Organic Chemistry at the University of California at Irvine, working under Dr. Robert

W. Taft. In 1970 he became an initiated disciple of His Divine Grace A. C. Bhaktivedanta Swami Prabhupāda, the founder-ācārya of the International Society for Krishna Consciousness. He worked as a Research Fellow at Emory University from 1974 to 1977, and is now Director of Bhaktivedanta Institute.

Dr. Singh has written a book, *The Scientific Basis of Krishna Consciousness,* and is co-author of several technical papers:

"Novel Activation Parameters and Catalytic Constants in the Aminolysis and Methanolysis of p-Nitrophenyl Trifluoro-acetate," *J. Amer. Chem. Soc.,* **97,** 3867 (1975).

"Nitrogen to Nitrogen Proton Transfer. The Significance of Large Negative Entropies of Activation," *J. Amer. Chem. Soc.,* **98,** 5011 (1976).

"Application of Ion Cyclotron Resonance Spectroscopic Gas-Phase Basicities to the Study of Tautomeric Equilibria," *J. Amer. Chem. Soc.,* **98,** 6048 (1976).

He is a member of the American Chemical Society and the International Society for the Study of the Origin of Life.

Dr. Michael Marchetti
(Mādhava Dāsa)

Dr. Marchetti was born in Orange, New Jersey, in 1943. At Rutgers University he received his A.B. and was elected a member of Sigma Xi, the scientific honorary society. He received his Ph.D. in Theoretical Chemistry from Georgetown University in 1970. He then worked at the National Bureau of Standards on a National Science Foundation Fellowship. In 1972 he took initiation from His Divine Grace, A. C. Bhakti-vedanta Swami Prabhupāda. As a member of Bhaktivedanta Institute he has presented courses and lectures at many universities in the Eastern U.S.A.

Dr. Marchetti has published several technical papers, including:

"Theoretical $^1\Sigma_g^+ - {}^1\Sigma_u^-$ Dipole Strengths of some Homonuclear Diatomic Molecules: Configuration Interaction," *J. Chem. Phys.*, **48**, 434 (1968).

"Electronic Transition Moment of the Lyman Bands of H_2 ($B^1\Sigma_u^+ - X^1\Sigma_g^+$) as a Function of Internuclear Separation," *J. Chem. Phys.*, **55**, 1665 (1971).

He is a member of Sigma Xi, the American Chemical Society, and the American Physical Society.

Dr. Richard Thompson
(Sadāputa Dāsa)
    Dr. Thompson was born in Binghamton, New York, in 1947. He received his B.S. in Mathematics and Physics from the State University of New York at Binghamton, and his M.A. in Mathematics from Syracuse University. In 1974 he received his Ph.D. in Probability Theory from Cornell University. He has worked for General Aniline and Film Corp. and Computer Science Corp. as a mathematician and computer programmer. In 1975 he became an initiated disciple of His Divine Grace, A. C. Bhaktivedanta Swami Prabhupāda. He is presently a member of Bhaktivedanta Institute.
    Dr. Thompson has published a monograph on the theory of statistical mechanics:

Equilibrium States on Thin Energy Shells, Memoirs of the American Mathematical Society, No. 150, 1974.

He is a member of the American Mathematical Society.

Robert S. Cohen
(Brahmatīrtha Dāsa)

Mr. Cohen was born in Newark, New Jersey, in 1949. He received his B.S. in Chemistry from Rensselaer Polytechnic Institute in 1971, and his M.S. in Geology from Rutgers University in 1975. He is presently employed as an exploration geologist. In 1976 he took initiation as a disciple of His Divine Grace, A. C. Bhaktivedanta Swami Prabhupāda.

Mr. Cohen is co-author of a book, *Perfect Questions — Perfect Answers*. He is also co-author of several technical articles, including:

"m-Benzene Disulfonic Acid (BDS) as a Superior Accompanying Acid for Routine Silicate Rock Analysis," *Chem. Geol.*, 16, 307 (1975).

He is a member of the Geological Society of America, the National Association of Geology Teachers, and the Society of Economic Paleontologists and Mineralogists.

To make lecture engagements, please write to Bhaktivedanta Institute at one of the following addresses:

70 Commonwealth Avenue
Boston, Massachusetts 02116
U.S.A.
Telephone: (617) 266-8369

Hare Krishna Land
Juhu, Bombay 400 054
India
Telephone: 57-9373

Bhaktivedanta Gurukula and
Institute for Higher Studies
Bhaktivedanta Swami Marg
Vrindavana, Mathura
India

FIRST INTERNATIONAL SCIENTIFIC CONFERENCE

on

LIFE COMES FROM LIFE

Sponsored by:

BHAKTIVEDANTA INSTITUTE

Founder — Acarya

His Divine Grace A. C. Bhaktivedanta Swami Prabhupada

MAIN SPEAKERS

Thoudam Damodar Singh. Manipur, India. Ph.D. in physical organic chemistry, University of California. GBC, and Director of Bhaktivedanta Institute.

Michael Marchetti. Boston, U.S.A. Ph.D. in theoretical chemistry, Georgetown University.

Richard Thompson. Boston, U.S.A. Ph.D. in mathematics, Cornell University.

David John Webb. London, England. M.A. in natural sciences, Oxford University.

Robert Cohen. Houston, U.S.A. M.A. in geology, Rutgers University.

HIGHLIGHTS OF THE CONFERENCE

● Molecular Biology

The goal of scientific research is to find the absolute truth or cause of all phenomena, governing both life and matter. According to modern science, the ultimate cause is vaguely incorporated into certain physical laws — basically the laws of quantum mechanics. Conceptually, these laws involve only some pushes and pulls among particles. Although the theory of evolution asserts that these laws are sufficient in themselves to account for all the marvels of life, honest and intelligent scientists are beginning to realize that the theory is failing to explain many subtle aspects of life — for instance, love, meaning and purpose. How can simple pushes and pulls be responsible for all the wonderful phenomena that we encounter in life?

We would like to show that the recent announcement of Khorana's synthetic gene is not different from that of Wohler's synthesis of urea in 1828, as far as our understanding of life is concerned. In other words, a complex molecule or a combination of such molecules will not account for the true nature of life.

We propose that life is non-physical and non-chemical. It possesses consciousness and obeys higher order nonphysical laws. This new scientific hypothesis can explain all the observed facts of life more scientifically than any of the previous theories.

● Information Theory

A fundamental proposition of information theory states that the information content of a closed mathematical system cannot increase. In modern science, nature is described by means of mathematical models of low information content. The physical structures of living organisms, on the other hand, are of such complexity and diversity as to indicate a very high information content. From this it follows that these structures could not have arisen by means of the simple natural processes envisioned in the theory of evolution. An additional source of information is required. The implications of this analysis regarding the origin of life are discussed.

● Quantum theory

Modern science has failed to account for consciousness, and in modern physics the existence of consciousness has given rise to serious paradoxes and contradictions. The quantum theory is discussed with special emphasis on the interpretations of von Neumann and Daneri, Loinger, and Prosperi. It is concluded that a new theoretical understanding is required containing provisions for consciousness. Such an understanding may be based on the description of the atomic living entity, or *atma,* given in *Bhagavad-gita.* This entails definite implications concerning the scientific study of life.

Current scientific theory holds that "life" is a product of molecular interactions, and that all the different activities of life — for example, thinking, feeling and willing — are due to coordinated chemical reactions. However, this approach has failed to explain the subtle aspects of life, such as value, meaning and purpose. A new scientific paradigm is, therefore, needed.

Bhaktivedanta Institute is sponsoring a scientific conference on the theme, "Life Comes from Life," to be held in Vrindavana, India on Friday, October 14 through Sunday, October 16. Scientists will gather from around the world to discuss fundamental questions on the nature and origin of life. The basic theme of the conference is that life cannot be reduced to a com-

PROGRAM

Friday, October 14.
7-9 AM   Breakfast
9-12 AM   Morning Session:
The Fundamental Difference Between Life and Matter.
12-2 PM   Lunch
2-5 PM   Afternoon Session:
Quantum Theory and the Laws of Consciousness.
6-8 PM   Dinner

Saturday, October 15.
7-9 AM   Breakfast
9-12 AM   Morning Session:
Demonstration by Information Theory that Life Cannot Arise from Matter.
12-2 PM   Lunch
2-5 PM   Afternoon Session:
Chemical Evolution — A Molecular Fairy Tale?
6-8 PM   Dinner

Sunday, October 16.
7-9 AM   Breakfast
9-12 AM   Morning Session:
Darwin's Theory and the Past History of Life.
12-2 PM   Lunch
2-5 PM   Afternoon Session:
Scientific Basis for the Study of Life.
6-8 PM   Dinner

Each session begins with a short address by one of the chief guests of the conference. Then the lecture for the session is given, followed by an open discussion period dealing with the lecture and related topics.

SETTING

The conference will be held in the historic holy city of Vrindavana, India. Vrindavana is located approximately 150 km southeast of Delhi and 50 km northwest of Agra. Lodging is provided in the International Guesthouse, situated next to the Krishna-Balaram Mandir, one of the most beautiful temples in Vrindavana. The meetings will be held next door in the newly opened Bhaktivedanta Gurukula and Institute for Higher Studies.

Meals are provided in the guesthouse restaurant. The menu features a sumptuous 10 to 15 course selection of traditional vegetarian dishes.

At this time of year the climate in Vrindavana is extremely pleasing. The sunshine during the daytime is quite mild. The surrounding neem trees are decorated with lustrous green leaves, and the dancing peacocks make constant appearances among the branches, thus pleasing the eyes greatly. The mild and cool breeze in the night carries the sweet sounds of kartals and bells, filling the atmosphere with a mood of serenity conducive to the minds of the seekers of real scientific knowledge.

bination of material elements. This will be discussed in the light of modern scientific theories and the Vedic knowledge. The members of the Institute hope that these discussions will provide new and stimulating insights into these important questions.

I will be able to attend the First International Scientific Conference on *Life Comes From Life*. I expect to stay at the Krishna-Balaram International Guesthouse for:

☐ 1 day   ☐ 2 days   ☐ 3 days

I will need transportation between the conference site and: ☐ Delhi   or   ☐ Agra.

Name _____

Address _____

City _____ Telephone _____

Please send the completed registration form to Dr. Onkar Nath Sharma, the secretary of the conference, at the following address:

Bhaktivedanta Gurukula and Institute for Higher Studies
c/o Krishna-Balaram Mandir, Raman Reti
Bhaktivedanta Swami Marg
Vrindavana, Mathura
India

For all invited guests, food, lodging, and transportation between Delhi and Agra and the conference site will be arranged free of charge.

List of speakers, professional backgrounds (partial) at time of conference:

Bhatt, S. R., Ph.D. Associate Professor of Philosophy at Delhi University. Author of *Studies in Ramanuja Vedanta* (1975) and *The Philosophy of Pancharatra: Advaitic Approach* (1968). Studied for a time at the Institute of Oriental Studies, Vrindavan, India, 1960s.

Cohen, Robert, M.S. in Geology, Rutgers University (1975). B.S. in Chemistry, Rensselaer Polytechnic Institute (1971). Exploration Geologist with Conoco Oil Company, Houston, TX. Co-author of technical papers in *Chemical Geology*, v. 15, 16.

Kirpal, Prem Nath, Ph.D. Chairman of the Executive Board of UNESCO (1970–72). Secretary, Government of India Ministry of Education (1960–68). Professor of History and Political Science, University of Lahore.

Malviya, A. N., Ph.D. Professor Department of Biochemistry, S. N. Medical College, Agra, U.P., India.

Marchetti, Michael, Ph.D. in Theoretical Chemistry, Georgetown University (1970). Worked at the National Bureau of Standards on a National Science Foundation Fellowship. Author of technical papers in the *Journal of Chemical Physics*, v. 48, 55.

Mishra, R. K., Ph.D. Professor & Head, Department of Biophysics, All India Institute of Medical Sciences, Delhi.

Ramaiah, A., Ph.D. Professor of Biochemistry, All India Institute of Medical Sciences, Delhi.

Singh, Thoudam Damodara, Ph.D. in Physical Organic Chemistry, University of California, Irvine (1974). Director of Bhaktivedanta Institute. Research Fellow at Emory University (1974–77). Author of technical papers in *Journal of the American Chemical Society*, v. 97, 98.

Soni, B. K., Ph.D. Deputy Director General of Animal Science Division, Indian Council of Agricultural Research.

Thompson, Richard, Ph.D. in Probability Theory and Statistical Mechanics, Cornell University (1974). M.A. in Mathematics, Syracuse University. Worked at Computer Science Corp., Silver Springs, MD, as a mathematician and computer programmer. Author of the monograph, *Equilibrium States on Thin Energy Shells*, Memoirs of the American Mathematical Society, No. 150.

Webb, David, J., M.A. in Chemistry, Wadham College, University of Oxford. Postgraduate Certificate of Education, University of London.

FIRST INTERNATIONAL "LIFE COMES FROM LIFE" CONFERENCE

FRIDAY, OCTOBER 14

Time	Duration	Topic	Speaker
9-30 AM	50 minutes	Breakfast	
10-30 AM	10 minutes	Welcoming Speech	Dr. T. D. Singh
10-40 AM	10 minutes	"Life and its Purpose"	Dr. P. N. Kripal
11-00 AM	1½ hours	"Life and matter"	Dr. T. D. Singh
12-30 PM	30 minutes	Questions and Discussion	
1-00 PM	2 hours	Lunch and break	
3-00 PM	2 hours including discussion	"Darwin's Theory and the past History of Life	Mr. R. Cohen
5-00 PM	1½ hours including discussion	"Philosophical Foundations of Science"	Dr. M. Marchetti
7-00 PM	1½ hours	Prasadam	
8-30 PM		Films	

SATURDAY, OCTOBER 15

Time	Duration	Topic	Speaker
7-45 AM	1 hour	Breakfast	
9-00 AM	30 minutes	"Philosophical Foundations of Life"	Dr. Bhatt
9-30 AM	10-15 minutes	Discussion	
9-45 AM	1¼ hours	"Demonstration by Information Theory that Life cannot Arise from Matter"	Dr. R. Thompson
11-15 AM	30 minutes	Discussion	
12-00 AM	2 hours	Lunch Break	
2-00 PM	1¼ hours	"Thermodynamics and the theory of Chemical Evolution"	Mr. D. J. Webb
3-30 PM	30 minutes	Discussion	
4-00 PM	20 Minutes	"Limitations of Science and Scientific Methods"	Dr. Ramayah
4-20 PM	1 hour	"Chemical Evolution"	Dr. T. D. Singh
5-00 PM	40 minutes	Informal Discussion	
7-00 PM	1¼ hours	Prasadam	
8-30 PM		Films and Slides	

SUNDAY, OCTOBER 16

Time	Duration	Topic	Speaker
7-45 AM	1 hour	Breakfast	
9-00 AM	30 minutes including discussion	'Life and its Value"	Dr. Soni
9-30 AM	30 minutes including discussion	"Theory of Living States"	Dr. Misra
10-00 AM	2 hours including discussion	"Quantum Mechanics and the Laws of Consciousness"	Dr. R. Thompson
12-00 AM	2 hours	Lunch Break	
2-00 PM	20 minutes including discussion	"The Process of Energy Transfer (Cytochrome c)"	Dr. Malviya
2-20 PM	1 hour	Summary of Conference and 'Scientific Basis for the Study of Life'	Dr. T. D. Singh
3-20 PM	1 hour 10 minutes	Discussion	
4-30 PM	15 minutes	Conclusion	Dr. T. D. Singh
4-45 PM	15 minutes	Vote of Thanks	

*This newspaper article about the conference is taken from
a recording of the report being read during a follow-up meeting.*

"The Nonphysical View on the Origin of Species"

The Statesman (Calcutta) – October 22, 1977

Materialists and men of faith continue to disagree over the origins of life. According to the first group, life is derived from atoms and molecules. The Russian scientist Dr. A. I. Oparin has been propagating this view since 1957, but his challengers demand really solid examples of life arising from matter. At a three-day international conference on "Life Comes from Life" at Vrindavan last week at the Bhaktivedanta Institute, it was stressed that life was independent of matter and dependent on higher principles lying beyond the present limitations of physics and chemistry.

The assumption that life itself was nonphysical was the keynote. The conference was opened by Dr. Prem Kirpal, former president of the executive board of UNESCO. Three lectures were delivered by Dr. Thoudam D. Singh, director of the Institute; Mr. Robert Cohen, a geologist from the USA; and Dr. Michael Marchetti, a theoretical chemist and student of the philosophy of science, on the fundamental nature of life and matter, new findings in paleontology and their effect on the theory of evolution, and the social consequences of a materialistic view of life. The philosophical foundations of life was the theme of a discourse by Dr. S. R. Bhatt, Associate Professor of Philosophy at Delhi University. Dr. Richard Thompson, a mathematician from Cornell University, and Mr. David Webb from England dealt with the application of information theory to the theory of evolution, thermodynamics and the origin of life. The limitations of science were discussed by Dr. A. Ramaiah, Professor of Biochemistry at the All India Institute of Medical Science.

Dr. Singh opposed the theory that life could be understood solely in terms of chemical combinations. There were intricate features of life, ranging from the structure of molecules in living cells to the subtle ones of human personality. The simple push-pull laws of chemistry and physics cannot account for these phenomena, and life and matter are understood as two distinct kinds of energy.

Mr. Cohen said that proof of the Darwinian theory of evolution must depend in the end on the fossil record. Darwin's theory required that all the different species of life were gradually transformed, one into another, through many small changes (mutations). Yet prominent paleontologists such as Eldridge and Gould are now maintaining that the fossil record only supports the view that species remain static in form and that changes between them, if they do really occur at all, can only occur by abrupt leaps. An examination of possible causes for such leaps shows that they could only be accounted for by the action of a higher intelligence, he said.

Dr. Thompson dealt with the mathematical analysis of the laws of nature studied in modern chemistry and physics. In the light of the modern theory of information, these laws can be shown to be unable to account for the highly complex and unique structures of living organisms. It can also be shown that the quantum mechanical laws suffer from serious shortcomings, because they cannot account for the nature of any conscious observer. Both of these lines of evidence supported the view that the living being is a nonphysical entity and that the behavior of matter when in the presence of life proves that there must be further higher-order laws and principles as yet unknown to modern science.

All of these conclusions were in agreement with the observed phenomena of life, and they also corroborate the systematic description of the nature of life given in *Bhagavad-gītā*. There was a general agreement among the participants of the conference that this approach to understanding the nature of life provided a viable alternative to the materialistic view of modern science.

Every Town and Village

A look at the worldwide activities of the International Society for Krishna Consciousness.

ISKCON Scientists Disclose Life's Origin

In mid-October, as the world's attention turned to the 1977 Nobel Prize science awards in Stockholm, a group of scientists met in the holy town of Vṛndāvana, India, to begin changing the direction of modern scientific research. The First International "Life Comes From Life" Conference, sponsored by ISKCON's Bhaktivedanta Institute, drew government and academic scientists from around the world. Also present was His Divine Grace A. C. Bhaktivedanta Swami Prabhupāda, the founder-*ācārya* of both ISKCON and the Institute.

Meanwhile, in Stockholm, Russian-born Ilya Prigogine was receiving an award for his thermodynamic mathematical models, which other scientists have tried to use in their as yet fruitless attempts to prove that life comes from chemical combinations. (Despite Prigogine's models, no scientist has ever observed life coming from chemical combinations, either in nature or in the laboratory. Nevertheless, Prigogine received $145,000 for his work.) On the other hand, the Bhaktivedanta Institute members at the "Life Comes From Life" Conference in Vṛndāvana conclusively proved that life can't possibly come from chemicals and that—as we see daily—life comes from life.

The members of the Bhaktivedanta Institute describe themselves as "a body of scientists and scholars who have recognized the unique value of the teachings of Kṛṣṇa consciousness brought to the West by His Divine Grace A. C. Bhaktivedanta Swami Prabhupāda." "One reason for the increasing interest of modern intellectuals in Śrīla Prabhupāda's teachings," they explain, "is doubtlessly the growing awareness that despite great scientific and technological advancement, the real goal of human life has somehow been missed. The philosophy of the Bhaktivedanta Institute provides a meaningful answer to this concern by proposing that life—not matter—is the basis of the world we perceive." The members add, "The Institute is clearly demonstrating that the Vedic philosophy is not a matter simply of 'faith' or 'belief' but is scientific in the strict sense of the term."

The "Life Comes From Life" Con-

Speakers at the "Life Comes From Life" Conference: (foreground) Dr. Thoudam Damodar Singh and Dr. Michael Marchetti; (rear) D. J. Webb, Robert S. Cohen, and Dr. Richard Thompson.

ference took place in the pleasant surroundings of ISKCON's modern Kṛṣṇa-Balarāma temple complex, with its blend of traditional Indian architecture and Western conveniences. Guests stayed at the temple's International Guest House and dined on delicious vegetarian fare from the guest house restaurant, all free of charge. Mild sunshine during the day and cool breezes at night made for a serene atmosphere.

The main speakers at the conference were Dr. Thoudam Damodar Singh (Svarūpa Dāmodara dāsa) and Dr. Richard Thompson (Sadāpūta dāsa), both of the Bhaktivedanta Institute. Dr. Singh, the Institute's director, was born in Manipur, India, in 1941 and became a disciple of Śrīla Prabhupāda in 1970. He has written a book entitled *The Scientific Basis of Kṛṣṇa Consciousness* and holds a Ph.D. in physical organic chemistry from the University of California.

To support the conclusion that life comes from life, Dr. Singh presented his findings in the field of molecular biology. Modern science is based on quantum mechanics, which reduces physical phenomena to pushes and pulls among certain particles. Having made these assumptions, scientists go on to say that life and life symptoms result from complex combinations of molecules. But Dr. Singh showed clearly that such theories

do not adequately explain the varied phenomena of life (including thought, emotions, and will). Life is based on consciousness, he proposed, and this consciousness obeys higher-order non-physical laws imposed by the supreme consciousness described in the *Vedas*.

Born in 1947 in Binghamton, New York, Dr. Thompson received his Ph.D. in probability theory from Cornell University in 1974. In 1975 he became an initiated disciple of Śrīla Prabhupāda. At the conference he delivered a well-received address called "Demonstration by Information Theory that Life Cannot Arise from Matter." A fundamental proposition of information theory states that the information content of a closed mathematical system cannot increase. Now, modern science describes physical nature in terms of mathematical models of low information content, and yet the physical structures of living organisms are so complex that they indicate a very high information content. So, according to information theory, it is impossible to suppose that life's high-information structures can arise from physical nature's low-information structures. On this basis Dr. Thompson discredited the now widely accepted theory that life forms do in fact evolve spontaneously from the ingredients of physical nature. Such evolution requires an outside source of information—namely, the supremely conscious controller of matter described in the *Vedas*.

Among other Bhaktivedanta Institute members who spoke was Dr. Michael Marchetti (Mādhava dāsa), who lectured on the philosophical foundations of science. Geologist Robert S. Cohen (Brahmatīrtha dāsa) demonstrated that fossil records actually give little support to Darwin's theory of evolution. Finally, Oxford's D.J. Webb (Jñāna dāsa) demonstrated how the laws of thermodynamics contradict current theories of chemical evolution.

At the end of the Vṛndāvana conference Dr. Singh announced that similar conferences will take place soon in Europe and North America. For further information about the Bhaktivedanta Institute, readers may write to 70 Commonwealth Avenue, Boston, Massachusetts 02116, or Hare Kṛṣṇa Land, Gandhi Gram Road, Juhu, Bombay 400 054, India.

*Members of the Monograph Series editorial board reviewing the
project. From left: Svarūpa Dāmodara Dāsa (Thoudam D. Singh),
Ravīndra Svarūpa Dāsa (William Deadwyler), Sadāpūta Dāsa
(Richard L. Thompson), and Rupānuga Dāsa (Robert Corens).*

NOTE: The copy of the Bhaktivedanta Institute document
on the opposite page, signed by the charter members of the institute as its
board of trustees, was donated from the files of Robert F. Corens.

ISKCON

INTERNATIONAL SOCIETY FOR KRISHNA CONSCIOUSNESS

FOUNDER-ACHARYA: HIS DIVINE GRACE A. C. BHAKTIVEDANTA SWAMI PRABHUPADA

10310 OAKLYN ROAD, POTOMAC, MARYLAND 20854

PHONE: 299-2100

We, the undersigned, hereby form The Bhaktivedanta Institute
as ordered by His Divine Grace A.C. Bhaktivedanta Swami
Prabhupada, Founder-Acarya of The International Society
for Krsna Consciousness. As charter members of the Institute
we assume responsibility for the management of the Institute
as its Board of Trustees under the direction of His Divine
Grace. We humbly pray for his eternal guidance, and we aspire
to develop the Institute as a glorification of His Divine
Grace, the bonafide representative of the Supreme Personality
of Godhead Lord Sri Krsna, and world teacher of genuine
scientific knowledge.

Dated this tenth day of December 1976, the Auspicious Appearance
Day of His Divine Grace ~~A.C.~~ Srila Bhaktisiddhanta Sarasvati Thakur.

Dr. Swarup Damodara das
(Thoudam D. Singh, Ph.D.)

Dr. Sadaputa das
(Richard Thompson, Ph.D.)

Dr. Rabindra Swarup das
(William H. Deadwyler, Ph.D.)

Dr. Madhava das
(Michael Marchetti, Ph.D.)

Rupanuga das
(Robert F. Corens)

APPROVED: _____

His Divine Grace A.C. Bhaktivedanta Swami

www.ingramcontent.com/pod-product-compliance
Lightning Source LLC
Chambersburg PA
CBHW051205200326
41519CB00025B/7012